SpringerBriefs in Electrical and Computer Engineering

For further volumes:
http://www.springer.com/series/10059

Dietmar Kissinger

Millimeter-Wave Receiver Concepts for 77 GHz Automotive Radar in Silicon-Germanium Technology

 Springer

Dietmar Kissinger
University of Erlangen-Nuremberg
Erlangen, Germany

ISSN 2191-8112   e-ISSN 2191-8120
ISBN 978-1-4614-2289-1   e-ISBN 978-1-4614-2290-7
DOI 10.1007/978-1-4614-2290-7
Springer New York Dordrecht Heidelberg London

Library of Congress Control Number: 2012932208

Printed on acid-free paper

Springer is part of Springer Science+Business Media (www.springer.com)

For Claudia

Your support allowed me to complete this work without losing sight of life.

Preface

In recent years the continuous improvement of silicon technologies lead to the realization of transistors with cut-off frequencies beyond 200 GHz. This development has enabled the integration of cost-efficient transceivers in the millimeter-wave regime that profit from high integration densities and yield of silicon-based technologies. Applications for such systems span from automotive radar to high data rate communication and high resolution imaging.

Automotive radar systems are considered the key technology for the realization of active and passive vehicular safety features to further reduce the number of fatalities due to traffic accidents. Silicon-Germanium technology offers the possibility of cost-efficient manufacturing of such systems for a broad range of traffic participants.

This book presents the analysis and design of integrated automotive radar receivers in Silicon-Germanium technology for use in complex multi-channel radar transceiver front-ends in the 77 GHz frequency band. The main emphasis of my work is the realization of high-linearity and low-power modular receive channels as well as the investigation of millimeter-wave integrated receiver test concepts.

The manuscript is based on my Ph.D. dissertation entitled "High-Linearity Circuits and Integrated Test Concepts for 77-GHz Radar Receiver Front-Ends in Silicon-Germanium Technology" that has been submitted to the University of Erlangen-Nuremberg, Germany. It has been supported by the German Bundesministerium für Bildung und Forschung (BMBF) through the research project RoCC - Radar on Chip for Cars under contract number 13N9821 and was carried out in close collaboration with Danube Integrated Circuit Engineering (DICE) GmbH, Linz, Austria and Infineon Technologies AG, Neubiberg, Germany.

Erlangen   Dietmar Kissinger

Acknowledgements

First and foremost I would like to thank my research advisor Prof. Dr.-Ing. Dr.-Ing. habil. Robert Weigel for giving me the opportunity to develop this work under his supervision. His guidance and support have been a constant encouragement for me to proceed with this topic over the past 4 years without hesitation.

I would like to thank Florian Starzer and Martin Jahn, as well as Dr. Christoph Wagner from the automotive radar group at DICE for a scientifically fruitful time and a warm welcome during my first years in Linz. The support of Dr. Linus Maurer, Dr. Erich Kolmhofer, and Dr. Herbert Jäger is also highly appreciated. Thomas Niedermayr, Florian Dober, and Andreas Schinko helped me with various layout issues and constantly kept me in a very positive work environment. Also, I would like to express my gratitude to Hans Peter Forstner, Dr. Herbert Knapp, Dr. Klaus Aufinger, and Dr. Rudolf Lachner from Infineon Technologies in Munich.

A special thanks goes to all my colleagues from the Institute for Electronics Engineering. I believe the research environment here in Erlangen is a very unique one, and I hope that it will continue for many years to come. I want to express my special gratitude to Christoph Kandziora, Jochen Rascher, Jochen Eßel, Dr. Alexander Kölpin, and Dr. Benjamin Waldmann for their motivation and mental support. Also, I thank my former students Roman Agethen, Abhiram Chakraborty, Timo Schürer, Johannes Pohlmann, and Andreas Selz for all their hard work and patience.

Furthermore, I sincerely thank my friends Andreas Schmidt, Denis Müller, and Sabine Rynek for providing me with blithely times outside my work place and a lot of rememberable coffee discussions.

I am very thankful to my parents and to my brother Thomas for their support and encouragement for my decision to continue my time at the university and develop this book. Finally, I would like to apologize to all my friends and relatives for my lack of time during the last years and promise a change for the better.

Contents

Acronyms

AC	Alternating current
ACC	Adaptive cruise control
ADC	Analog-to-digital converter
Balun	Balanced-to-unbalanced converter
BiCMOS	Bipolar complementary metal oxide semiconductor
BITE	Built-in test equipment
BPF	Bandpass filter
CB	Common-base
CE	Common-emitter
CG	Common-gate
CMOS	Complementary metal oxide semiconductor
CS	Common-source
CW	Continuous wave
CUT	Circuit under test
DAC	Digital-to-analog converter
DC	Direct current
DDS	Direct digital synthesis
DSB	Double-sideband
DSP	Digital signal processor
DUT	Device under test
EIRP	Equivalent isotropic radiated power
EM	Electromagnetic
ENR	Excess noise ratio
FFT	Fast Fourier transformation
FMCW	Frequency-modulated continuous wave
FOM	Figure of merit
HBT	Heterojunction bipolar transistor
HS	High speed
HV	High voltage
IC	Integrated circuit
IF	Intermediate frequency

IQ	In-phase and quadrature
IRR	Image rejection ratio
LFSR	Linear feedback shift register
Lidar	Light detection and ranging
LNA	Low-noise amplifier
LO	Local oscillator
LPF	Low-pass filter
LRR	Long range radar
LSB	Lower sideband
MIM	Metal-insulator-metal
MRR	Mid range radar
NEP	Noise equivalent power
NF	Noise figure
PA	Power amplifier
PCB	Printed circuit board
PLL	Phase-locked loop
PRBS	Pseudo-random binary sequence
Radar	Radio detection and ranging
RCS	Radar cross section
RF	Radio frequency
RFIC	Radio frequency integrated circuit
RMS	Root mean square
ROM	Read-only memory
RX	Receive
SAW	Surface acoustic wave
SE	Single-ended
SiGe	Silicon-germanium
SiGe:C	Silicon-germanium: carbon
SiP	System in package
SNR	Signal-to-noise ratio
SoC	System on chip
SRR	Short range radar
SSB	Single-sideband
TEM	Transmission electron microscopy
TX	Transmit
UHS	Ultra-high speed
UWB	Ultra-wide band
USB	Upper sideband
VCO	Voltage-controlled oscillator
VNA	Vector network analyzer

Chapter 1
Introduction

1.1 Motivation

Since its birth the evolution of microelectronics following Moore's law has led to ever rising integration densities with constantly decreasing feature sizes [1]. The resulting reduction of costs per transistor has laid the foundation for the success of the semiconductor industry and the constant improvement of product performance and functionality. Silicon-based technologies offer high yield in combination with high integration levels which makes them suitable for cost-efficient high volume manufacturing for mass market applications. The key device in highly-integrated microwave transceiver front-ends is the silicon-germanium heterojunction bipolar transistors. The heterostructure principle has been proposed in detail as early as 1957 by Kroemer [2]. Nevertheless the necessity for precise epitaxial technologies such as molecular beam epitaxy and chemical vapor deposition delayed the first reported integration until 1987 [3]. Kroemer received the Nobel Prize in Physics 2000 "for basic work on information and communication technology".

In recent years the constant improvement of silicon-germanium process technology with reported maximum oscillation frequencies and cutoff frequencies well above 200 GHz [4–9] has enabled the full integration of transceiver front-ends operating in the millimeter-wave regime for a variety of applications such as automotive radar, high data rate short-range communication, and passive imaging [10]. Automotive radar systems are considered the key technology to realize active and passive vehicular safety functions. Traditionally, expensive III–V semiconductor chipsets and discrete components have been used to implement front-ends in the millimeter-wave domain. Silicon-germanium technology based integrated transceivers offer the potential to manufacture cost-efficient radar sensors for a broad range of traffic participants. Figure 1.1 shows a third generation adaptive cruise control (ACC) radar sensor from Robert Bosch GmbH. It was introduced in 2009 and features the first silicon-germanium based fully integrated 77 GHz four-channel transceiver. Future automotive radar systems will incorporate a high

D. Kissinger, *Millimeter-Wave Receiver Concepts for 77 GHz Automotive Radar in Silicon-Germanium Technology*, SpringerBriefs in Electrical and Computer Engineering, DOI 10.1007/978-1-4614-2290-7_1, © Springer Science+Business Media, LLC 2012

a   b

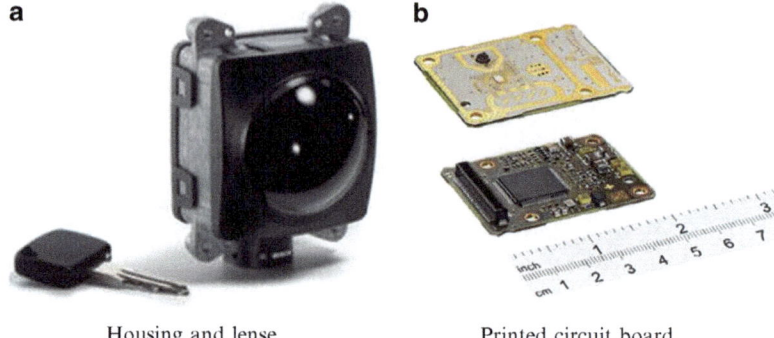

Housing and lense   Printed circuit board

Fig. 1.1 Third generation adaptive cruise control (ACC) radar sensor from Robert Bosch GmbH featuring the first silicon-germanium based fully integrated 77 GHz transceiver

number of transmit and receive paths to facilitate beam-steering and phased array functionality. Among the key parameters toward a successful implementation of such systems is the realization of low-power receivers featuring high-linearity performance.

After manufacturing millimeter-wave front-ends have to be tested for correct functionality and performance before they are passed on to the consumer. Testing of radio frequency (RF) building blocks and fully integrated systems at millimeter-wave frequencies necessitates high-performance measurement equipment and careful calibration procedures which is expensive both in terms of time and cost. This task becomes even more challenging for future highly complex multi-channel front-ends and increasing operational frequencies. For single-chip integrated systems this issue is further complicated by the lack of access to signal nodes between the individual high-frequency building blocks. The expenses for testing millimeter-wave integrated circuits is approaching a significant part of the overall manufacturing cost and presents a stumbling block when trying to reduce costs of radar sensors and address today's time-to-market demands.

Built-In Test Equipment (BITE) offers the possibility to simplify testing procedures through simple low frequency off-chip interfaces at the expense of an increased chip area and power consumption for the additional testing circuitry [11]. Furthermore, an on-line diagnosis provides the possibility to constantly monitor the high-frequency performance parameters and functionality of the transceiver. These features are especially useful in safety-relevant applications like automotive radar systems. The introduction of high-frequency silicon-germanium technologies with their high level of integration has created opportunities for such cost-effective embedded test solutions.

1.2   Current State of the Art

The feasibility of silicon-germanium (SiGe) technologies for 77 GHz automotive radar applications was first proven in 2004 by Li and Rein [12]. They presented a fully integrated SiGe voltage-controlled oscillator (VCO) with high output power necessary for vehicular radar applications. The oscillator featured a wide tuning range required to account for process tolerances and the specified temperature range. The design was further improved by Pohl et al. for 79 GHz ultra-wide bandwidth short-range applications [13]. Several other groups have also reported the successful generation of high output powers above 10 dBm in the 77 GHz frequency range in SiGe technology [14–16].

The first SiGe based 77 GHz down-conversion mixers were published in 2004 by Perndl et al. [17] and in 2006 by Dehlink et al. [18]. The authors presented active mixer topologies resembling a standard double-balanced Gilbert cell approach. A modification of the above architecture for single-ended input signals known as the micromixer topology has also been presented [19,20]. Receiver front-ends featuring additional low-noise amplifier (LNA) stages prior to the mixer have been published by different groups [21–25]. Furthermore, investigations of stand-alone LNAs have been performed [26–29]. Following publications of an integrated transceiver in 2006 [30, 31] the first fully qualified single-chip 77 GHz radar transceiver in SiGe has been presented in 2008 by Forstner et al. [32]. To date a number of successful integrations of monolithic 77 GHz SiGe transceivers have been reported [24,33–35].

1.3   Goals of this Work

Future automotive radar systems for vehicular safety purposes demand high sensitivities for an improved target detection. In monostatic direct-conversion frequency-modulated continuous wave (FMCW) automotive radar systems this necessitates the implementation of a low-noise stage prior to the down-conversion stage to lower the noise figure of the overall system and reduce unwanted emissions. Additionally, the large dynamic range of the received signal requires high-linearity performance of the receiver with input compression points in the range of -15 to -5 dBm due to strong interferers and reflected signals, caused by bumper reflections and mismatch of the transmitting antenna. Furthermore, the trend toward radar systems with a high number of channels for phased-array operation demands low-power operation of the receiver channels. State-of-the-art published receivers suffer either from poor linearity [17,21–25,30] or low sensitivity and high power consumption [18–20,32]. Derived from these facts, the first goal of this work is the investigation and design of receiver front-ends with the above specifications of simultaneous high sensitivity and dynamic range as well as low-power consumption. Moreover, the design has to comply with the common voltage levels of the overall automotive radar transceiver system to facilitate easy and modular integration.

Prior published work in the field of integrated test equipment mainly focused on mixed-signal systems [36–38]. At higher frequencies the loop-back approach has been proposed for integrated transceivers which gives only limited insight regarding the performance and possible failures of the individual building blocks [39]. To date integrated test solutions at millimeter-wave frequencies do not exist. The second goal of this work is the analysis and design of potential built-in test circuitry for integrated 77 GHz automotive radar receiver front-ends in SiGe technology. Furthermore, these solutions have to keep additional efforts in terms of chip area and power consumption at a minimum.

1.4   Organization of the Book

After the introduction, the following chapter gives an overview of radar fundamentals including antenna concepts and different carrier modulation schemes. Applications of automotive radar technology as well as current international frequency regulations are outlined. Chapter 3 introduces the silicon-germanium process used throughout this work. It includes a review of the implemented transistor models and an overview of the active and passive devices utilized in this technology. In Chap. 4 millimeter-wave receiver concepts are described. The mixer principle as a core component of the receiver is outlined and different receiver topologies are discussed. The chapter concludes with an outline of relevant mixer performance parameters and the introduction of the differential circuit design technique. Chapter 5 presents a differential 77 GHz high-linearity receiver front-end. In comparison to other published work, the circuit incorporates a reduced voltage mixer architecture and shows the highest dynamic range to power consumption ratio as necessary for the realization of integrated multi-channel radar transceivers. In Chap. 6 a novel differential current re-use low-noise amplifier architecture with improved isolation for high-linearity receivers is introduced. The broadband design shows the highest gain per stage, the highest gain-bandwidth product, as well as the highest output-referred 1-dB compression point at comparable power consumption. Chapter 7 introduces millimeter-wave built-in test concepts. Approaches for test signal generation, coupling, and signal detection for direct built-in test architectures are outlined. In Chap. 8 an in-depth analysis of different direct-conversion receiver built-in test architectures suitable for 77 GHz automotive radar systems is performed. Chapter 9 presents a novel recursive 77 GHz mixer test architecture. The proposed mixer architecture is capable of simultaneous up- and down-conversion to enable a functionality test of the receiver path. Furthermore, the presented design proves especially useful for complex multi-channel integrated receiver front-ends. The book concludes with Chap. 10 that summarizes the results and provides an outlook.

References

1. G. E. Moore, "Cramming more components onto integrated circuits," *Electronics*, vol. 38, no. 8, pp. 114–117, Aug. 1965.
2. H. Kroemer, "Theory of a wide-gap emitter for transistors," *Proc. IRE*, vol. 45, no. 11, pp. 1535–1537, Nov. 1957.
3. S. S. Iyer, G. L. Patton, S. S. Delage, S. Tiwari, and J. M. C. Stork, "Silicon-germanium base heterojunction bipolar transistors by molecular beam epitaxy," in *IEEE Int. Electron Devices Meeting Tech. Dig.*, Washington, D.C., Dec. 1987, pp. 874–876.
4. J. Böck, H. Schäfer, H. Knapp, K. Aufinger, M. Wurzer, S. Boguth, T. Böttner, R. Stengl, W. Perndl, and T. F. Meister, "3.3 ps SiGe bipolar technology," in *IEEE Int. Electron Devices Meeting Tech. Dig.*, San Francisco, CA, Dec. 2004, pp. 255–258.
5. R. K. Vytla, T. F. Meister, K. Aufinger, D. Lukashevich, S. Boguth, J. Böck, H. Schäfer, and R. Lachner, "Simultaneous integration of SiGe high speed transistors and high voltage transistors," in *Proc. Bipolar/BiCMOS Circuits Technol. Meeting*, Maastricht, The Netherlands, Oct. 2006.
6. B. A. Orner, M. Dahlström, A. Pothiawala, R. M. Rassel, Q. Liu, H. Ding, M. Khater, D. Ahlgren, A. Joseph, and J. Dunn, "A BiCMOS technology featuring a 300/330 GHz (f_T/f_{max}) SiGe HBT for millimeter wave applications," in *Proc. Bipolar/BiCMOS Circuits Technol. Meeting*, Maastricht, The Netherlands, Oct. 2006.
7. H. Rücker, B. Heinemann, R. Barth, J. Bauer, K. Blum, D. Bolze, J. Drews, G. G. Fischer, A. Fox, O. Fursenko, T. Grabolla, U. Haak, W. Höppner, D. Knoll, K. Köpke, B. Kuck, A. Mai, S. Marschmeyer, T. Morgenstern, H. H. Richter, P. Schley, D. Schmidt, K. Schulz, B. Tillack, G. Weidner, W. Winkler, D. Wolansky, H.-E. Wulf, and Y. Yamamoto, "SiGe BiCMOS technology with 3.0 ps gate delay," in *IEEE Int. Electron Devices Meeting Tech. Dig.*, Washington, D.C., Dec. 2007, pp. 651–654.
8. G. Avenier, P. Chevalier, G. Troillard, B. Vandelle, F. Brossard, L. Depoyan, M. Buczko, S. Boret, S. Montusclat, A. Margain, S. Pruvost, S. T. Nicolson, K. H. K. Yau, D. Gloria, D. Dutartre, S. P. Voinigescu, and A. Chantre, "0.13 μm SiGe BiCMOS technology for mm-wave applications," in *Proc. Bipolar/BiCMOS Circuits Technol. Meeting*, Monterey, CA, Oct. 2008, pp. 89–92.
9. E. Preisler, L. Lanzerotti, P. D. Hurwitz, and M. Racanelli, "Demonstration of a 270 GHz f_T SiGe-C HBT within a manufacturing-proven 0.18 μm BiCMOS process without the use of a raised extrinsic base," in *Proc. Bipolar/BiCMOS Circuits Technol. Meeting*, Monterey, CA, Oct. 2008, pp. 125–128.
10. E. Kasper, D. Kissinger, P. Russer, and R. Weigel, "High speeds in a single chip," *IEEE Microw. Mag.*, vol. 10, no. 7, pp. 28–33, Dec. 2009.
11. D. Kissinger, B. Laemmle, L. Maurer, and R. Weigel, "Integrated test for silicon front ends," *IEEE Microw. Mag.*, vol. 11, no. 3, pp. 87–94, May 2010.
12. H. Li, H.-M. Rein, T. Suttorp, and J. Böck, "Fully integrated SiGe VCOs with powerful output buffer for 77-GHz automotive radar systems and applications around 100 GHz," *IEEE J. Solid-State Circuits*, vol. 39, no. 10, pp. 1650–1658, Oct. 2004.
13. N. Pohl, H.-M. Rein, T. Musch, K. Aufinger, and J. Hausner, "SiGe bipolar VCO with ultra-wide tuning range at 80 GHz center frequency," *IEEE J. Solid-State Circuits*, vol. 44, no. 10, pp. 2655–2662, Oct. 2009.
14. U. R. Pfeiffer, S. K. Reynolds, and B. A. Floyd, "A 77 GHz SiGe power amplifier for potential applications in automotive radar systems," in *Proc. IEEE Radio Frequency Integr. Circuits Symp.*, Fort Worth, TX, Jun. 2004, pp. 91–94.
15. A. Komijani and A. Hajimiri, "A wideband 77-GHz, 17.5-dBm fully integrated power amplifier in silicon," *IEEE J. Solid-State Circuits*, vol. 41, no. 8, pp. 1749–1756, Aug. 2006.
16. A. Ghazinour, P. Wennekers, R. Reuter, Y. Yi, H. Li, T. Böhm, and D. Jahn, "An integrated SiGe-BiCMOS low noise transmitter chip with a frequency divider chain for 77 GHz applications," in *Proc. Eur. Microw. Integr. Circuits Conf.*, Manchester, United Kingdom, Sep. 2006, pp. 194–197.

17. W. Perndl, H. Knapp, M. Wurzer, K. Aufinger, T. F. Meister, J. Böck, W. Simbürger, and A. L. Scholtz, "A low-noise and high-gain double-balanced mixer for 77 GHz automotive radar front-ends in SiGe bipolar technology," in *Proc. IEEE Radio Frequency Integr. Circuits Symp.*, Fort Worth, TX, Jun. 2004, pp. 47–50.

18. B. Dehlink, H.-D. Wohlmuth, H.-P. Forstner, H. Knapp, S. Trotta, K. Aufinger, T. F. Meister, J. Böck, and A. L. Scholtz, "A highly linear SiGe double-balanced mixer for 77 GHz automotive radar applications," in *Proc. IEEE Radio Frequency Integr. Circuits Symp.*, San Francisco, CA, Jun. 2006, pp. 235–238.

19. L. Wang, R. Kraemer, and J. Borngraeber, "An improved highly-linear low-power down-conversion micromixer for 77 GHz automotive radar in SiGe technology," in *IEEE MTT-S Int. Microw. Symp. Dig.*, San Francisco, CA, Jun. 2006, pp. 1834–1837.

20. M. Hartmann, C. Wagner, K. Seemann, J. Platz, H. Jäger, and R. Weigel, "A low-power micromixer with high linearity for automotive radar at 77 GHz in silicon-germanium bipolar technology," in *IEEE Topical Meeting on Silicon Monolithic Integr. Circuits in RF Syst. Dig.*, Long Beach, CA, Jan. 2007, pp. 237–240.

21. ——, "A low-power low-noise single-chip receiver front-end for automotive radar at 77 GHz in silicon-germanium bipolar technology," in *Proc. IEEE Radio Frequency Integr. Circuits Symp.*, Honolulu, HI, Jun. 2007, pp. 149–152.

22. R. Reuter, H. Li, I. To, Y. Yin, A. Ghazinour, D. Jahn, D. Morgan, J. Feige, P. Welch, S. Braithwaite, B. Knappenberger, D. Scheitlin, J. John, M. Huang, P. Wennekers, M. Tutt, C. Trigas, and J. Kirchgessner, "Fully integrated SiGe-BiCMOS receiver(RX) and transmitter(TX) chips for 76.5 GHz FMCW automotive radar systems including demonstrator board design," in *IEEE MTT-S Int. Microw. Symp. Dig.*, Honolulu, HI, Jun. 2007, pp. 1307–1310.

23. L. Wang, S. Glisic, J. Borngräber, W. Winkler, and J. C. Scheytt, "A single-ended fully integrated SiGe 77/79 GHz receiver for automotive radar," *IEEE J. Solid-State Circuits*, vol. 43, no. 9, pp. 1897–1908, Sep. 2008.

24. S. T. Nicolson, P. Chevalier, B. Sautreuil, and S. P. Voinigescu, "Single-chip W-band SiGe HBT transceivers and receivers for Doppler radar and millimeter-wave imaging," *IEEE J. Solid-State Circuits*, vol. 43, no. 10, pp. 2206–2217, Oct. 2008.

25. J. Powell, H. Kim, and C. G. Sodini, "SiGe receiver front ends for millimeter-wave passive imaging," *IEEE Trans. Microw. Theory Tech.*, vol. 56, no. 11, pp. 2416–2425, Nov. 2008.

26. B. A. Floyd, "V-band and W-band SiGe bipolar low-noise amplifiers and voltage-controlled oscillators," in *Proc. IEEE Radio Frequency Integr. Circuits Symp.*, Fort Worth, TX, Jun. 2004, pp. 295–298.

27. B. Dehlink, H.-D. Wohlmuth, K. Aufinger, T. F. Meister, J. Böck, and A. L. Scholtz, "A low-noise amplifier at 77 GHz in SiGe:C bipolar technology," in *IEEE Compound Semicond. Integr. Circuits Symp. Tech. Dig.*, Palm Springs, CA, Nov. 2005, pp. 287–290.

28. R. Reuter and Y. Yin, "A 77 GHz (W-band) SiGe LNA with 6.2 dB noise figure and gain adjustable to 33 dB," in *Proc. Bipolar/BiCMOS Circuits Technol. Meeting*, Maastricht, The Netherlands, Oct. 2006.

29. S. Chartier, B. Schleicher, F. Korndörfer, S. Glisic, G. Fischer, and H. Schuhmacher, "A fully integrated fully differential low-noise amplifier for short range automotive radar using a SiGe:C BiCMOS technology," in *Proc. Eur. Microw. Integr. Circuits Conf.*, Munich, Germany, Oct. 2007, pp. 407–410.

30. A. Babakhani, X. Guan, A. Komijani, A. Natarajan, and A. Hajimiri, "A 77 GHz phased-array transceiver with on-chip antennas in silicon: Receiver and antennas," *IEEE J. Solid-State Circuits*, vol. 41, no. 12, pp. 2795–2806, Dec. 2006.

31. A. Natarajan, A. Komijani, X. Guan, A. Babakhani, and A. Hajimiri, "A 77 GHz phased-array transceiver with on-chip antennas in silicon: Transmitter and local LO-path phase shifting," *IEEE J. Solid-State Circuits*, vol. 41, no. 12, pp. 2807–2819, Dec. 2006.

32. H.-P. Forstner, H. Knapp, H. Jäger, E. Kolmhofer, J. Platz, F. Starzer, M. Treml, A. Schinko, G. Birschkus, J. Böck, K. Aufinger, R. Lachner, T. F. Meister, H. Schäfer, D. Lukashevich, S. Boguth, A. Fischer, F. Reininger, L. Maurer, J. Minichshofer, and D. Steinbuch, "A 77GHz 4-channel automotive radar transceiver in SiGe," in *Proc. IEEE Radio Frequency Integr. Circuits Symp.*, Atlanta, GA, Jun. 2008, pp. 233–236.

33. V. Jain, F. Tzeng, L. Zhou, and P. Heydari, "A single-chip dual-band 22-29-GHz/77-81GHz BiCMOS transceiver for automotive radars," *IEEE J. Solid-State Circuits*, vol. 44, no. 12, pp. 3469–3485, Dec. 2009.

34. S. Trotta, B. Dehlink, A. Ghazinour, D. Morgan, and J. John, "A 77 GHz 3.3 V 4-channel transceiver in SiGe BiCMOS technology," in *Proc. Bipolar/BiCMOS Circuits Technol. Meeting*, Capri, Italy, Oct. 2009, pp. 186–189.

35. F. Starzer, H. P. Forstner, C. Wagner, A. Fischer, H. Jäger, D. Kissinger, and A. Stelzer, "A 77-GHz FMCW radar transceiver sourced through a 19-GHz SiGe colpitts oscillator," in *Proc. Asia-Pacific Microw. Conf.*, Singapore, Dec. 2009, pp. 2304–2307.

36. G. W. Roberts, "Improving the testability of mixed-signal integrated circuits," in *Proc. IEEE Custom Integr. Circuits Conf.*, Santa Clara, CA, May 1997, pp. 214–221.

37. B. Dufort and G. W. Roberts, "On-chip analog signal generation for mixed-signal built-in self-test," *IEEE J. Solid-State Circuits*, vol. 34, no. 3, pp. 318–330, Mar. 1999.

38. S. S. Akbay, A. Halder, A. Chatterjee, and D. Keezer, "Low-cost test of embedded RF/analog/mixed-signal circuits in SOPs," *IEEE Trans. Adv. Packag.*, vol. 27, no. 2, pp. 352–363, May 2004.

39. J.-S. Yoon and W. R. Eisenstadt, "Embedded loopback test for RF ICs," *IEEE Trans. Instrum. Meas.*, vol. 54, no. 5, pp. 1715–1720, Oct. 2005.

Chapter 2
Radar Fundamentals

The abbreviation *radar* stands for "radio detection and ranging". It designates a radio technology for the determination of distances to remote stationary or moving objects. This chapter intends to give a brief overview about radar techniques [1].

2.1 Radar Equation

The radar principle is based on the properties of electromagnetic waves and its characteristic reflection at different materials. Thereby, a radio signal of wavelength λ is transmitted. Based on the reflected and received signal response conclusions regarding direction, distance, and relative velocity of the reflecting target can be drawn. The received signal strength of the target can be calculated after the radar equation:

$$P_r = \frac{P_t G_t A_r \sigma_S}{(4\pi)^2 R^4} \quad \text{with} \quad A_r = \frac{G_r \lambda^2}{4\pi}. \tag{2.1}$$

In the above expression, P_r denotes the received signal strength, while P_t represents the transmitted signal power. The antenna is characterized by its transmit and receive antenna gain G_t and G_r as well as the corresponding effective aperture A_r of the receiving antenna. σ_S is the scattering cross section of the reflecting target which is located at the distance R. The received signal strength degrades with a power of 4. This is in contrast to general communication systems which employ bidirectional transmission, leading to a much higher degradation in power levels. Therefore, the radar receiver has to provide a high sensitivity and dynamic range in order to cover a wide range of target distances.

D. Kissinger, *Millimeter-Wave Receiver Concepts for 77 GHz Automotive Radar in Silicon-Germanium Technology*, SpringerBriefs in Electrical and Computer Engineering, DOI 10.1007/978-1-4614-2290-7_2, © Springer Science+Business Media, LLC 2012

2.2   Antenna Concepts

One can distinguish between mono- and bistatic radar architectures. These differ in the design of the transmit and receive antennas. Bistatic radar devices possess spatially separated antennas for the transmit (TX) and receive (RX) path. Figure 2.1 shows a schematic diagram of a bistatic radar architecture [2].

In a monostatic radar architecture a single antenna performs both the transmission and reception of the radar signal. Figure 2.2 shows a schematic diagram of a monostatic radar transceiver. The transmitted and received signals are separated through a circulator. Signals generated in the transmitter are passed directly to the antenna, while received signals from the antenna are routed to the receiver part. An ideal circulator theoretically provides infinite isolation between the transmit and receive path. Nevertheless, the isolation of millimeter-wave circulators over a certain bandwidth is limited and coupling effects have to be accounted for.

2.3   Carrier Modulation

2.3.1   Pulse-Doppler Radar

The distance from a target to the radar system can be determined by measuring the propagation delay between the transmitted and received signal. Figure 2.3 shows a block diagram of a pulse-doppler radar architecture. The oscillator generates a radio

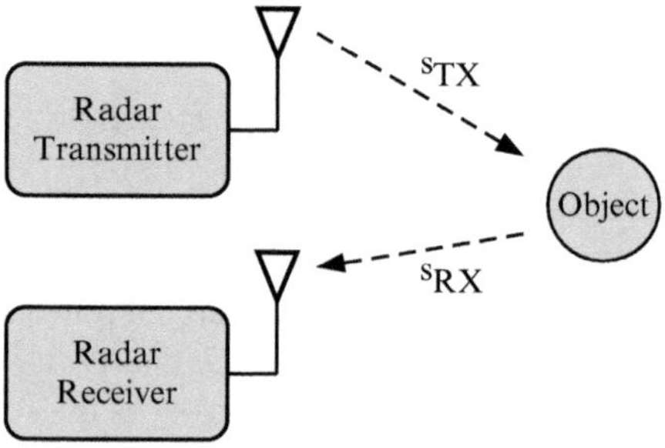

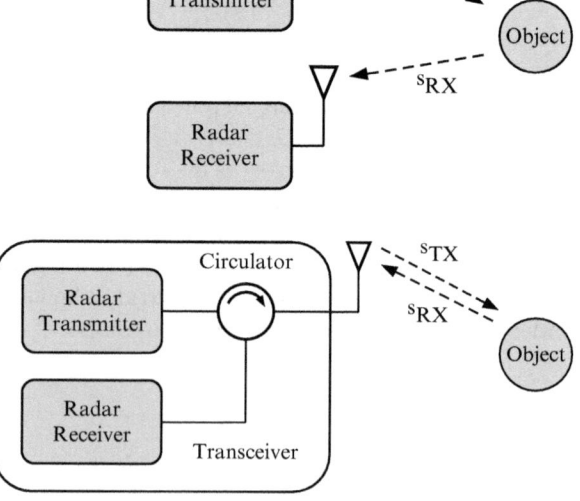

Fig. 2.1 Schematic diagram of a bistatic radar architecture consisting of spatially separated radar transmitter and receiver with dedicated antennas

Fig. 2.2 Schematic diagram of a monostatic radar architecture consisting of transmitter, receiver, and a circulator for signal splitting at the antenna

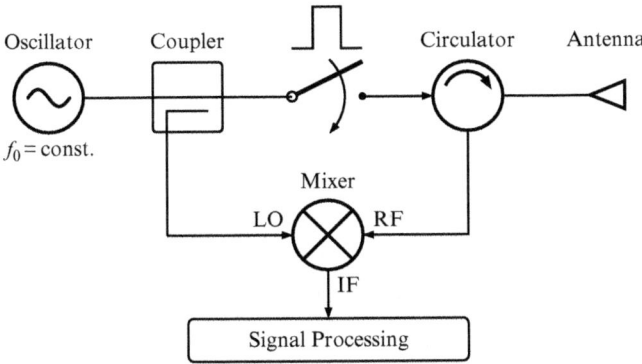

Fig. 2.3 Block diagram of a pulse-doppler radar transceiver architecture

signal at the constant frequency f_0 that is converted into a continuous pulse train by a pulse-shaping device and transmitted via the antenna. The incoming signal is mixed with the local oscillator (LO) signal and subsequently processed in the baseband.

If the propagation speed c of the electromagnetic wave in the medium is known, one can calculate the distance R between the obstacle and the radar system via the round trip propagation delay Δt of the impulse by (2.2). The radio wave propagation can be derived from the vacuum speed of light c_0 and material permittivity ε_r.

$$R = \frac{c\Delta t}{2} \quad \text{with} \quad c \approx \frac{c_0}{\sqrt{\varepsilon_r}} \tag{2.2}$$

In addition to the above determination of the range to the target, the relative velocity of the object with respect to the radar system can be derived from the Doppler shift of the received signal frequency f_t. Equation 2.3 gives the relationship between relative velocity v_r and the Doppler frequency shift $f_d = f_t - f_0$.

$$v_r = \frac{c f_d}{2 f_0} \quad \text{for} \quad v \ll c \tag{2.3}$$

The Doppler shift describes the shift of frequency caused by motion of the target with respect to the signal source. Depending on the direction of movement the shift is of positive value for objects approaching the signal source and takes a negative value for targets moving away from the radar transmitter.

Figure 2.4 shows the transmitted (a) and incoming (b) signals in a pulse-radar system. It possesses a maximum range which is defined by the pulse repetition rate T_P of the transmitter. The maximum unambiguous range R_{max} is given by

$$R_{\text{max}} = \frac{c T_p}{2} . \tag{2.4}$$

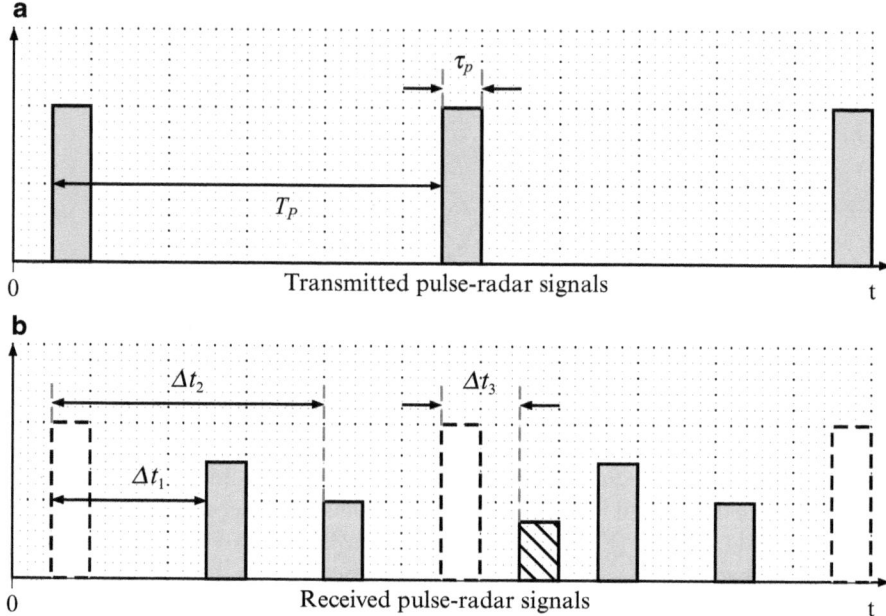

Fig. 2.4 Time dependent behavior of transmitted (**a**) and received (**b**) signals in a pulse-radar system; pulse repetition rate T_p, pulse width τ_p, propagation delay Δt

It defines the longest range a pulse can travel from the transmitter to the receiver before the next pulse is emitted. Violation of the above leads to the appearance of ghost targets as shown in Fig. 2.4b. In case the system receives an echo from a prior pulse after the following pulse has been transmitted, the distance to the target is falsely calculated to be Δt_3 instead of the actual range $T_p + \Delta t_3$. The range resolution of a pulse-radar dependent upon the pulse width τ_p is given as

$$\delta_r = \frac{c\,\tau_p}{2} \approx \frac{c}{2B}. \tag{2.5}$$

Two different targets can only be distinguished from each other as long as their individual backscattered pulse responses do not overlap and become blurred. The range resolution can be related to the inverse of the pulse bandwidth B by (2.5).

2.3.2 Frequency Modulated Continuous-Wave Radar

Unmodulated continuous-wave (CW) radars transmit a signal with constant frequency. The lack of modulation of the source only allows for determination of the relative target velocity via the Doppler shift. Frequency modulated continuous-wave

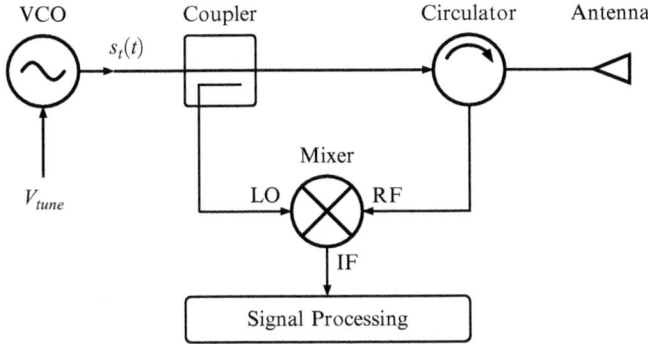

Fig. 2.5 Block diagram of a direct-conversion FMCW radar transceiver architecture

(FMCW) radar systems employ frequency modulation at the signal source to enable propagation delay measurements for determination of the distance to the target. Figure 2.5 shows a block diagram of an FMCW radar transceiver.

A voltage-controlled oscillator (VCO) forms the signal source that is modulated as a linear frequency ramp by changing the tuning voltage V_{tune} of the VCO. Equation 2.6 gives the mathematical expression for a frequency modulation that uses up- and down-chirps of equal length $T_P/2$ and bandwidth B.

$$f_t(t) = f_0 + kt \qquad \text{with} \quad k = \frac{2B}{T_P} \tag{2.6}$$

The above modulation scheme yields a transmitted signal s_t of the form

$$s_t(t) = A_t \cos\left(2\pi f_t(t)t\right) \tag{2.7}$$

$$= A_t \cos\left(2\pi f_0 t + 2\pi k t^2\right). \tag{2.8}$$

A fraction of the signal is coupled to the receive mixer to act as the LO reference, while the other part is transmitted through the antenna. The backscattered signal

$$s_r(t - \Delta t) = A_r \cos\left(2\pi(f_0 + f_d)(t - \Delta t) + 2\pi k(t - \Delta t)^2\right) \tag{2.9}$$

with the propagation delay Δt and a Doppler shift f_d is received and translated into the baseband by means of a down-conversion mixer. Subsequently the intermediate frequency (IF) signal is digitized and the determination of target range and velocity is performed through a fast Fourier transformation (FFT).

Figure 2.6a shows the variation in frequency versus time for the transmitted and received signal. The VCO generates an up- and down-chirp of bandwidth B within a modulation period T_P. After transmission the reflected signal is received with a propagation delay Δt. If the target exhibits a relative velocity to the transmitter

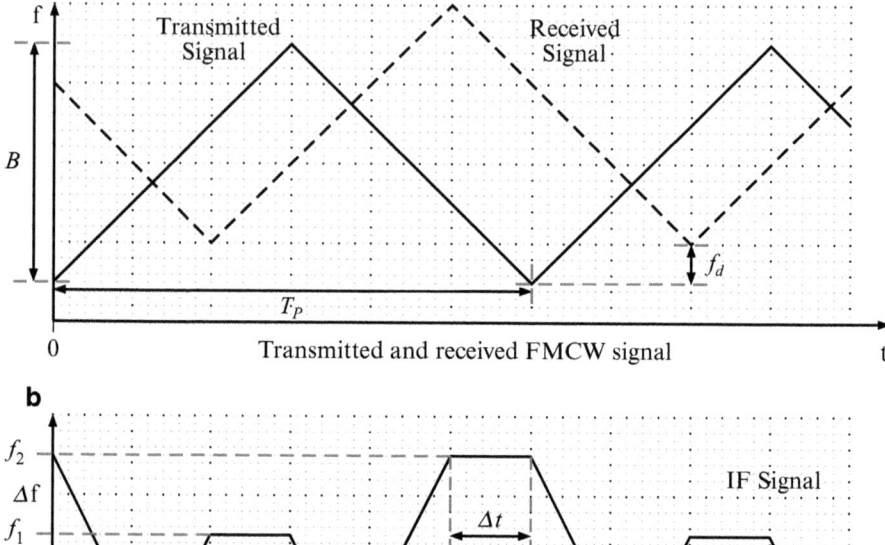

Fig. 2.6 Time dependent behavior of transmitted and received (**a**) and IF signal (**b**) of an FMCW radar system; propagation delay Δt, Doppler shift f_d, bandwidth B

the frequency of the received signal is shifted by f_d. The resulting IF signal is depicted in Fig. 2.6b. It represents the absolute of the frequency difference between the transmit and receive signal. Based on the two different IF frequencies f_1 and f_2 one can obtain the range R and relative velocity v_r of the target through

$$R = \frac{cT_p}{2B} \frac{f_1 + f_2}{2} \tag{2.10}$$

$$v_r = \frac{c}{2f_c} \frac{f_1 - f_2}{2}. \tag{2.11}$$

Furthermore, the range resolution δ_r of an FMCW radar system is given by

$$\delta_r = \frac{c}{2B}. \tag{2.12}$$

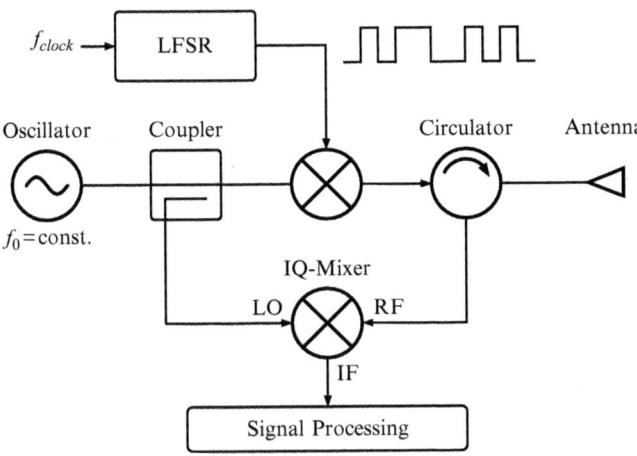

Fig. 2.7 Block diagram of a pseudo-noise modulated radar transceiver architecture

2.3.3 Pseudo-Noise Modulated Continuous-Wave Radar

In contrast to pulse-radar systems, where energy is transmitted in a short time period, the method of pulse compression allows to replace a pulse waveform by spread spectrum signals that distribute their energy over a long time. This can be achieved by phase or frequency modulation of the carrier with a pseudo-random binary sequence (PRBS) to reduce the peak power of the radar system [3]. A maximum length binary sequence (M-sequence) is a special type of PRBS signal that can be generated through a linear feedback shift register (LFSR). An N-stage shift register can generate a pseudo-random code of the length $2^N - 1$. Figure 2.7 shows a diagram of a PRBS modulated radar system.

The linear feedback shift register generates an M-sequence that is up-converted onto the carrier frequency f_0 by a biphase modulator [4]. The resulting double-sideband spectrum is transmitted via the antenna. To avoid loss of information, a quadrature down-conversion mixer is implemented in the receiver path, which translates the signal into the baseband [5]. The signal processing unit performs correlation of the IF signal and the reference PRBS code in the digital domain.

Figure 2.8 shows the exemplary time shape of a fourth-order M-sequence [6]. The pseudo random sequence is periodic with a period of T_p, that is related to the chip duration t_c and the number of bits N by

$$T_p = N t_c . \tag{2.13}$$

The power spectrum of an M-sequence is the Fourier transform of its autocorrelation function, which is approximately equal to a single rectangular segment. This results in a line spectrum as shown in Fig. 2.9 with an envelope shape $S(f)$ determined by a sinc^2 function according to (2.14). The spacing between the

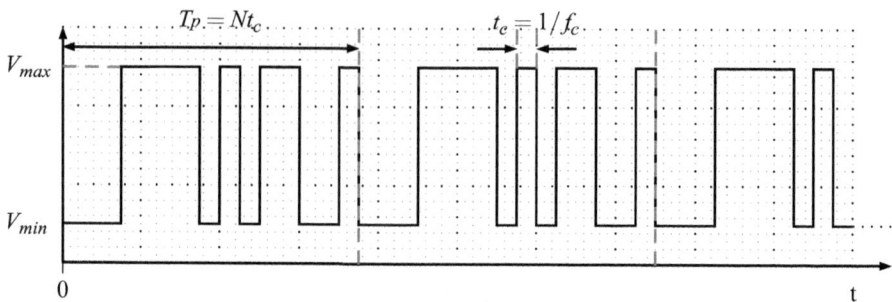

Fig. 2.8 Time shape of a fourth-order M-sequence; pseudo random sequence period T_p, number of bits N, chip duration t_c, and bandwidth f_c

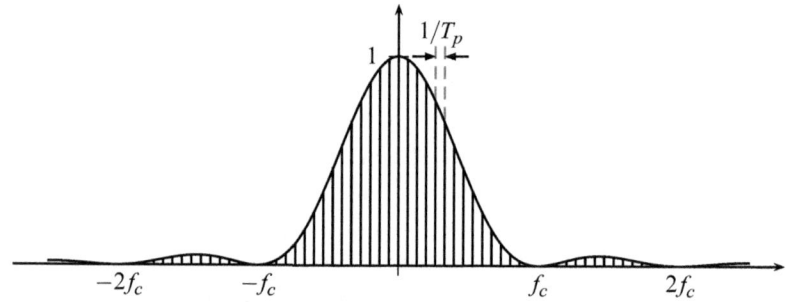

Fig. 2.9 Representation of a fourth-order M-sequence in the frequency domain; pseudo random sequence period T_p, chip duration t_c, and bandwidth f_c

individual lines depends on the period T_p of the M-sequence.

$$S(f) = \text{sinc}^2(\pi f t_c) = \frac{\sin^2(\pi f t_c)}{(\pi f t_c)^2} \tag{2.14}$$

A direct relationship between the bandwidth f_c of the PRBS code and the minimum achievable range resolution δ_r exists:

$$\delta_r = \frac{c}{2 f_c}, \tag{2.15}$$

and the supported unambiguous range $R_{\max}$ is calculated to

$$R_{\max} = \frac{c T_p}{2} = \frac{cN}{2 f_c}. \tag{2.16}$$

In order to achieve a low range resolution δ_r and at the same time support a large unambiguous range $R_{\max}$, the chip duration t_c is required to remain small while the bit-length N of the M-sequence has to be sufficiently large. Moreover, in ranging systems based on spread spectrum techniques, a large value of N is necessary to improve the autocorrelation function of the sequence which lowers the cross-correlation function. This allows the system to discriminate among other spread spectrum signals, e.g. radar transmitters of other traffic participants [7].

2.4 Automotive Radar

In Europe alone about 1.3 million traffic accidents cause more than 41,000 fatalities and an economical damage of more than 200 billion Euros per year. Human error is involved in over 90% of the overall accidents. The introduction of preventive safety applications into the car can avoid the above errors by e.g. helping the driver to maintain a safe speed and distance, keep within the lane, and prevent overtaking in critical situations. Figure 2.10 depicts a number of different technologies that are used in active safety systems to monitor the surrounding environment of a vehicle.

The first generation of driver assistance systems only featured comfort functions. Adaptive Cruise Control (ACC) based on 77 GHz Long Range Radar (LRR) automatically adjusts the driving speed of the vehicle to maintain a constant distance to the car ahead. Parking aid and parking slot measurement based on ultrasound and 24 GHz Short Range Radar (SRR) support the driver during the parking process.

In the current generation, passive safety features are employed. Lane change assistance and blind spot detection systems warn the driver if other vehicles would be critical in case of a lane changing maneuver. Pre-crash sensing and collision warning systems alert the driver of an imminent critical situation, pre-regulate the necessary braking pressure and prepare safety measures to mitigate the results of a

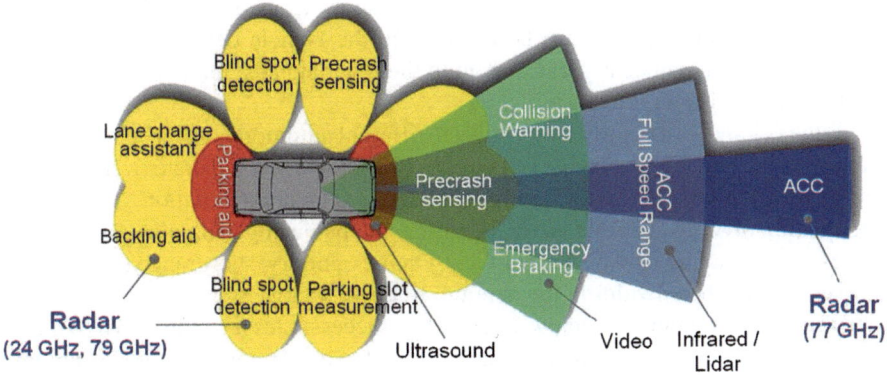

Fig. 2.10 Surrounding field monitoring technologies for driver assistance systems

Table 2.1 Comparison of different automotive radar sensor classifications

	Long range radar	Mid range radar	Short range radar
Frequency band	77 GHz	79 GHz	79 GHz
Max. output power EIRP	+55 dBm	−9 dBm/MHz	−9 dBm/MHz
Bandwidth	600 MHz	600 MHz	4 GHz
Distance range	10–250 m	1–100 m	0.15–30 m
Distance resolution	0.5 m	0.5 m	0.1 m
Speed resolution	0.1 m/s	0.1 m/s	0.1 m/s
Angular accuracy	0.1°	0.5°	1°
3 dB Beamwidth Azimuth	±15°	±40°	±80°
3 dB Beamwidth elevation	±5°	±5°	±10°

Table 2.2 Comparison of currently available automotive radar frequency bands

		24 GHz ISM[a]	24 GHz UWB[a]	77 GHz LRR	79 GHz SRR
Europe	Bandwidth	200 MHz	5 GHz[b]	1 GHz	4 GHz
	Power EIRP	+20 dBm	−41.3 dBm/MHz	+55 dBm	−9 dBm/MHz
USA	Bandwidth	100/250 MHz	7 GHz	1 GHz	Negotiations
	Power EIRP	+32.7/12.7 dBm	−41.3 dBm/MHz	+23 dBm	
Japan	Bandwidth	76 MHz	5 GHz	0.5 GHz	Planned
	Power EIRP	+10 dBm[c]	−41.3 dBm/MHz	+10 dBm[c]	

[a]Not protected for automotive radar
[b]Only until 2013
[c]Transmit power at antenna feed

possible collision. Future active safety systems will initiate an emergency brake if the sensor data indicates that a collision cannot be avoided. Additional rear sensors enable the determination of an optimum deceleration to avoid a potential collision with the following car [8].

In comparison to optical systems, e.g. video or lidar, radar based sensors can operate reliably under a variety of different environmental situations. They are robust against rough weather, e.g. rain, fog, or snow, as well as lighting conditions. Furthermore, radar sensors can be installed in the vehicle behind plastic radomes that are electromagnetically transparent to the radar signals, making them invisible in the exterior design of the car.

Table 2.1 shows a comparison between different automotive radar sensor classifications [9]. Long range radar sensors in the 76–77 GHz band operate over a distance range up to 200 m but require only a moderate distance resolution. Therefore, a bandwidth of only 200–600 MHz is sufficient, but a high angular resolution is necessary due to the narrow field of view. On the contrary, short range radar sensors possess a wide beamwidth in the azimuth to cover a large viewing area. The range accuracy requirements are below 10 cm which necessitates a bandwidth of 4 GHz.

Different frequency bands exist that are dedicated or can be used for automotive radar applications. Table 2.2 shows a comparison of the currently available frequency bands in the main markets. In Europe the 24 GHz UWB band has only been temporarily allocated for short range applications. From 2013 the frequency regulation demands that new cars have to be equipped with 79 GHz short range radar sensors [10]. Similar allocations are planned in Japan and are currently under negotiation in the Unites States.

References

1. M. I. Skolnik, *Radar Handbook*, 3rd ed. McGraw-Hill, 2008.
2. N. J. Willis and H. D. Griffiths, *Advances in Bistatic Radar*. Scitech Pub, 2007.
3. B. Sewiolo, "Ultra-wideband transmitters based on M-sequences for high resolution radar and sensing applications," Ph.D. dissertation, Inst. for Electron. Eng., Univ. of Erlangen-Nuremberg, Erlangen, Germany, 2010.
4. S. Trotta, H. Knapp, D. Dibra, K. Aufinger, T. F. Meister, J. Böck, W. Simbürger, and A. L. Scholtz, "A 79 GHz SiGe-bipolar spread-spectrum TX for automotive radar," in *IEEE Int. Solid-State Circuits Conf. Dig. Tech. Papers*, San Francisco, CA, Feb. 2007, pp. 430–431.
5. B. Dehlink, H.-D. Wohlmuth, K. Aufinger, F. Weiss, and A. L. Scholtz, "An 80 GHz SiGe quadrature receiver frontend," in *IEEE Compound Semicond. Integr. Circuits Symp. Tech. Dig.*, San Antonio, TX, Nov. 2006, pp. 197–200.
6. D. J. Daniels, *Ground Penetrating Radar*, 2nd ed. The Institution of Engineering and Technology, 2004.
7. H.-J. Zepernick and A. Finger, *Pseudo Random Signal Processing - Theory and Application*. John Wiley & Sons, 2005.
8. H. L. Bloecher, J. Dickmann, and M. Andres, "Automotive active safety & comfort functions using radar," in *IEEE Int. Conf. Ultra-Wideband*, Vancouver, Canada, Sep. 2009, pp. 490–494.
9. M. Köhler, J. Hasch, F. Gumbmann, L.-P. Schmidt, J. Schür, and H. L. Bloecher, "Automotive radar operation considerations and concepts at frequencies beyond 100 GHz," in *Proc. Eur. Radar Conf. Workshop "Automotive Radar Sensors in the 76-81 GHz Frequency Range"*, Paris, France, Oct. 2010.
10. K. M. Strohm, H. L. Bloecher, R. Schneider, and J. Wenger, "Development of future short range radar technology," in *Proc. Eur. Radar Conf.*, Paris, France, Oct. 2005, pp. 165–168.

Chapter 3
Silicon-Germanium Bipolar Technology

3.1 Bipolar Transistor Models

3.1.1 Static Large-Signal Behavior

The design of integrated circuits in the millimeter-wave frequency range requires compact and accurate transistor models for a fast and reliable simulation of the circuit performance prior to the manufacturing process. Transistor models can be subdivided into large-signal models and small-signal models [1]. While large signal models are mostly used for the determination of the direct current (DC) operating point of the transistor configuration, the latter are used for modeling the high-frequency behavior of the circuit being operated around a certain DC bias point. Figure 3.1 shows the schematic symbol with pin designation and an elementary large-signal transistor model of an npn bipolar transistor based on a modified version of the Ebers-Moll model [2].

The model in Fig. 3.1b employs two diodes D_{BE} and D_{BC} to describe the base-emitter and base-collector junctions of the npn transistor. A controlled current source between the collector and emitter is used to model the transfer current. The transistor is usually operated in the active forward region where the base-emitter diode is forward biased while the base-collector diode is in reverse mode. In this case the diode D_{BC} can be omitted and the bipolar transistor behaves like a current-controlled current source. A small base current I_B controls the collector current I_C via the forward current gain β_F,

$$I_C = \beta_F I_B . \tag{3.1}$$

The collector current I_C can be related to the base-emitter voltage V_{BE} as follows:

$$I_C = I_S \exp\left(\frac{V_{BE}}{V_T}\right)\left(1 + \frac{V_{CE}}{V_A}\right). \tag{3.2}$$

D. Kissinger, *Millimeter-Wave Receiver Concepts for 77 GHz Automotive Radar in Silicon-Germanium Technology*, SpringerBriefs in Electrical and Computer Engineering, DOI 10.1007/978-1-4614-2290-7_3, © Springer Science+Business Media, LLC 2012

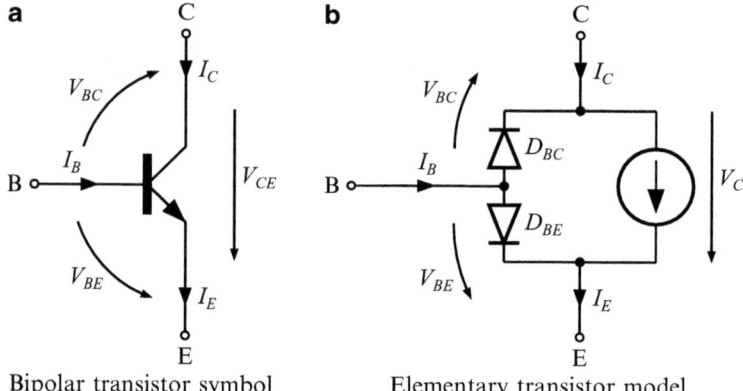

Fig. 3.1 Elementary model of an npn bipolar transistor. (**a**) Symbol with pin designators. (**b**) Equivalent circuit; Base B, Emitter E, and Collector C

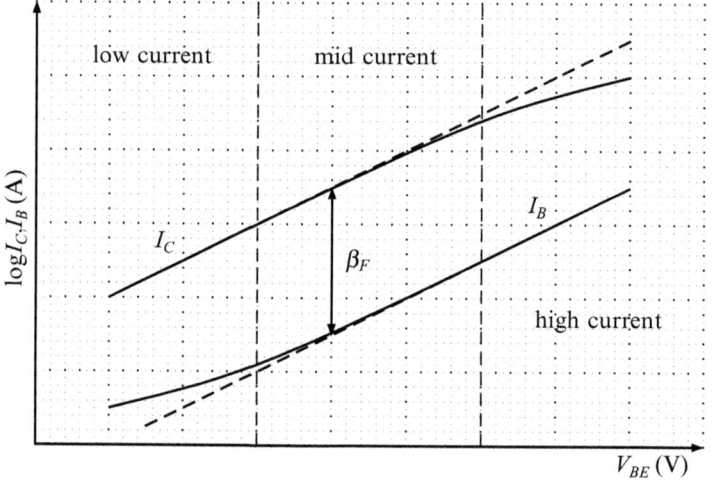

Fig. 3.2 Typical semi-logarithmic plot of collector I_C and base current I_B versus base-emitter voltage V_{BE} for a constant collector-emitter voltage V_{CE} (Gummel-Plot)

In (3.2) I_S denotes the transfer saturation current and V_A describes the Early voltage. The thermal voltage V_T is defined by (3.3) where k denotes the Boltzmann constant, T the temperature, and q the elementary charge.

$$V_T = \frac{kT}{q} \tag{3.3}$$

Figure 3.2 shows a typical semi-logarithmic plot of the collector and base currents I_C, I_B versus collector emitter voltage V_{CE} also known as a Gummel-Plot.

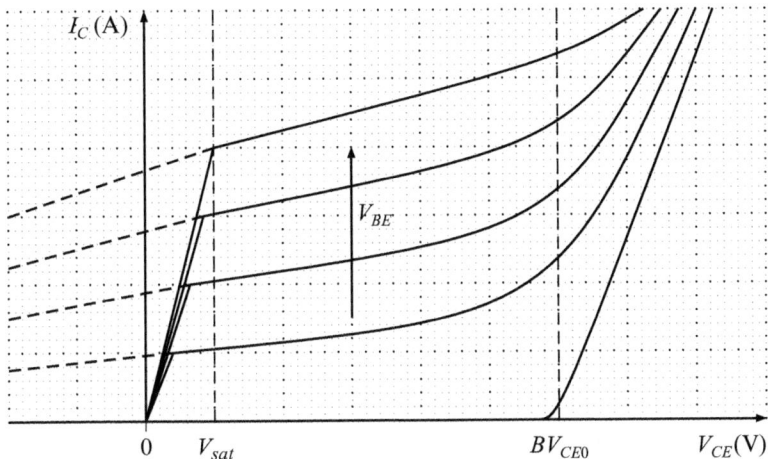

Fig. 3.3 Typical bipolar transistor transfer characteristic of collector current I_C versus collector-emitter voltage V_{CE} for different base-emitter voltages V_{BE}

The plot can be divided into three different regions, namely low, mid, and high current zone. The relationship in (3.1) only holds true for the mid current region where β_F resembles the constant difference between both lines [3].

In the low current zone the base current is dominated by recombination of charge carriers in the base-emitter depletion region that generate a leakage current.

In the high current region several effects occur that lead to collector current saturation. The emitter efficiency decreases as the base is flooded with minority carriers. This increases the base doping which in turn lowers the injection rate from the emitter into the base. A second effect is called emitter current crowding. Due to resistive voltage drops at the base, the base-emitter diode is more forward biased on the edges than in the center, leading to a reduced effective emitter area. Finally, the Kirk effect describes an increased effective base area, caused by the minor carrier concentration being close to the collector doping, which extends the base into the collector.

Figure 3.3 shows a typical transfer characteristic of a bipolar transistor. Operation in the forward active region is ensured for a collector-emitter voltage $V_{CE} \geq V_{sat}$. The collector current shows a dependency upon the collector-emitter voltage, which is expressed through the Early voltage V_A in (3.2). The Early voltage can be graphically depicted as the intersection point of the transfer functions with the voltage axis.

If a certain collector-emitter voltage is applied, the reverse biased base-collector diode is affected by avalanche breakdown. This leads to a strong increase of the collector current that can potentially damage the device. The maximum voltage level is typically specified by the open base breakdown voltage BV_{CE0} at $I_B = 0$. An important tradeoff exists between the cut-off frequency and the collector-emitter breakdown voltage of a transistor described by the Johnson limit.

Fig. 3.4 Static small-signal model of an npn transistor in the forward active region

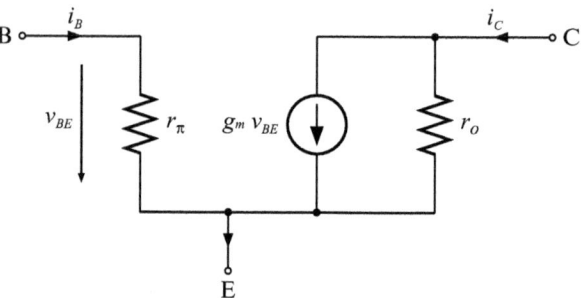

The above bipolar transistor model lacks several important effects that influence the dynamic large signal behavior of the transistor. The Gummel-Poon model represents an enhanced version which is based on the charge-control theory [4].

3.1.2 Small-Signal Parameters

A variety of circuit building blocks in a transceiver operate at small signal levels compared to the bias. One can derive a small-signal model through linearization around the operating point of the nonlinear Gummel-Poon model. Figure 3.4 shows the resulting static small-signal model of an npn bipolar transistor.

The model can be characterized by the following parameters:

$$g_m = \frac{\delta I_C}{\delta V_{BE}} = \frac{I_C}{V_T} \tag{3.4}$$

$$\beta_0 = \frac{\delta I_C}{\delta I_B} \tag{3.5}$$

$$r_\pi = \frac{\delta V_{BE}}{\delta I_B} = \frac{\beta_0}{g_m} \tag{3.6}$$

$$r_o = \frac{\delta V_{CE}}{\delta I_C} = \frac{V_A}{I_C}. \tag{3.7}$$

The input resistance r_π, the transconductance g_m, and the output resistance r_o can be determined as partial derivatives of the transfer characteristics of (3.1) and (3.2). The transconductance g_m describes the change of the collector current due to the change of the base-emitter voltage V_{BE} according to (3.4). The small-signal current gain β_0 is defined by (3.5) and shows the variation of the collector current I_C due to the change in the base current I_B. The input resistance r_π is calculated as a function of β_0 and g_m according to (3.6). The Early-effect is described through the output resistance r_o which depends on the Early voltage V_A and the collector current I_C.

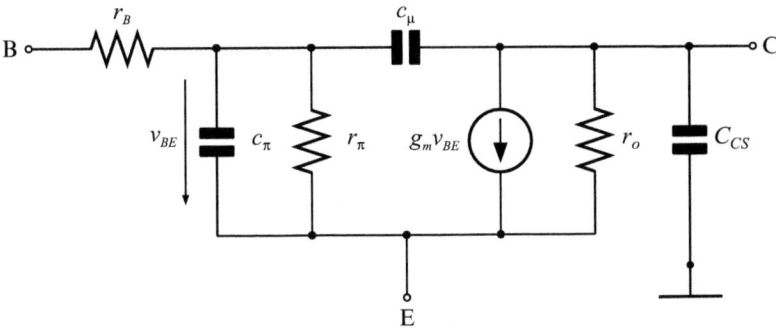

Fig. 3.5 High-frequency hybrid-π small-signal model of an npn transistor

The above model only holds true for low frequency small-signal analysis. Figure 3.5 shows the dynamic high-frequency hybrid-π model of an npn bipolar transistor which includes the base resistance as well as parasitic capacitances between the respective ports. These limit the high-frequency performance of the transistor.

Each of the two capacitances c_π and c_μ consist of a diffusion capacitance and a depletion part. In forward operation mode the diffusion capacitance c_μ of the reverse biased base-collector diode can be neglected, resulting in

$$c_\mu = c_{jc} \tag{3.8}$$

$$c_\pi = c_{je} + \tau_F g_m , \tag{3.9}$$

where c_{jc} and c_{je} denote the depletion capacitances of the base-collector and base-emitter diodes. The diffusion capacitance of the base-emitter diode is given through the product of the base forward transit time τ_F and the transconductance g_m and the frequency dependent small-signal current gain $\beta(j\omega)$ results in:

$$\beta(j\omega) = \frac{\beta_0}{1 + j\omega\beta_0 \frac{c_\pi + c_\mu}{g_m}} . \tag{3.10}$$

Different figures of merit to describe the high-frequency performance of transistors exist. The transit frequency f_t is a measure for the broadband behavior and switching speed of a transistor. It is defined as the frequency where the small-signal current gain $\beta(j\omega)$ drops to unity and is calculated after (3.11).

$$f_t = \frac{1}{2\pi} \frac{g_m}{c_\pi + c_\mu} \tag{3.11}$$

A second important figure of merit is the maximum frequency of oscillation f_{max} of a transistor which can be calculated as

$$f_{max} = \sqrt{\frac{f_t}{8\pi c_\mu r_B}} \,. \tag{3.12}$$

It is defined as the frequency where the unilateral power gain of the transistor becomes unity and is dependent upon the base resistance r_B.

3.2 SiGe:C Bipolar Technology

Over the last two decades silicon-germanium heterojunction bipolar transistors (SiGe HBT) have matured from laboratory research into mainstream radio frequency production technologies [5]. State of the art processes achieve transit and maximum frequencies of oscillation above 400 GHz [6]. The following sections introduce Infineon Technologies high performance silicon-germanium HBT technologies b7hf200 and b7hf500 which have been used throughout this work.

3.2.1 Active Devices

SiGe HBT technologies differ from regular bipolar transistors by their base doping and material [7]. In a SiGe HBT the base region is formed by an epitaxially grown silicon-germanium layer between the adjacent silicon layers. The lattice constants of silicon (Si) and silicon-germanium differ by about 4%. These SiGe films grown upon a Si substrate are compressively strained and subject to a stability criterion according to Fig. 3.6, which plots the realizable effective layer thickness against the strain, i.e. the germanium (Ge) content [8]. The bottom left green area marks the stable region in which defect-free SiGe films can be grown on Si substrates.

With increased germanium content in the base, the bandgap E_g of the base region is reduced by $\Delta E_{g,Ge}$, as shown in Fig. 3.7. Due to the lower bandgap of the base region and the high base doping a high emitter efficiency is achieved. The base of an HBT can thus be more heavily doped than the base of a Si bipolar transistor to reduce the base resistance without a negative impact on the current amplification. This, in combination with an accelerated drift field due to the gradient in the Ge base doping, increases the maximum frequency of oscillation f_{max}.

The process b7hf200 from Infineon Technologies is a fully qualified high performance bipolar SiGe:C technology intended for millimeter-wave applications, e.g. automotive radar and short range high data rate communication circuits [9, 10]. The technology provides three different npn transistor types and a lateral pnp transistor. The minimum feature size of the b7hf200 technology is 0.35 μm which results in an effective emitter width of 0.18 μm. Figure 3.8 shows a cross section of the npn HBT.

Fig. 3.6 Layer stability diagram of Si/SiGe heterostructures showing the realizable effective layer thickness versus the Ge content (strain) of the grown film for unconditionally stable defect-free SiGe layers on silicon

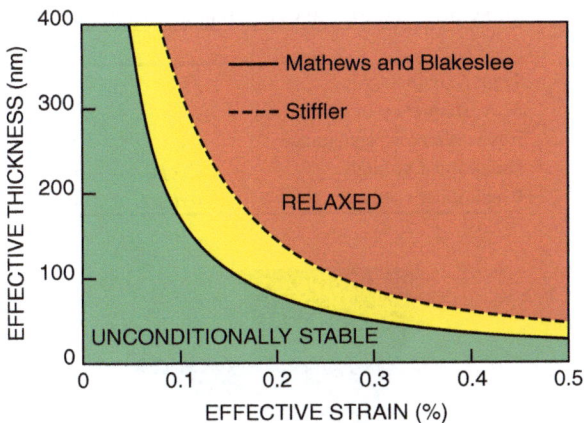

Fig. 3.7 Band structure of a SiGe heterojunction bipolar transistor. Due to the lower bandgap of the SiGe base region, a higher base doping with lower base resistance in comparison to Si bipolar transistors can be achieved

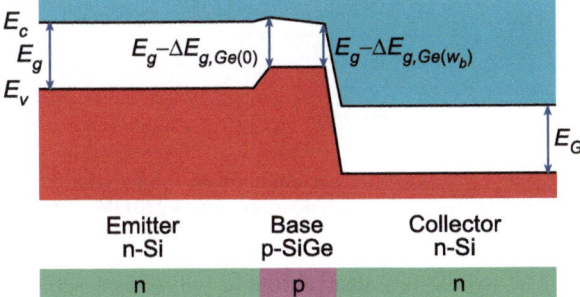

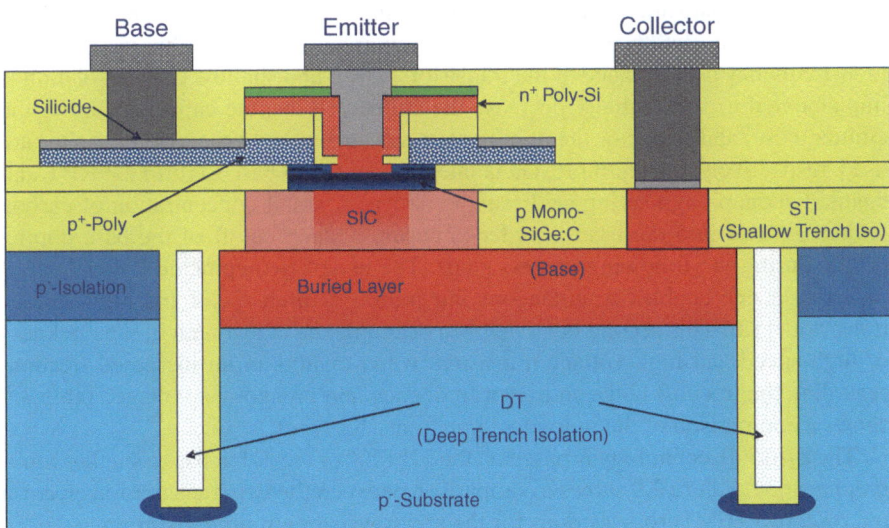

Fig. 3.8 Cross sectional view of a SiGe npn heterojunction bipolar transistor with self-implanted collector (SIC) and deep trench isolation in the b7hf200 technology

Table 3.1 Comparison of different transistor types available in b7hf200

		UHS npn	HS npn	HV npn
Transit frequency	f_T	200 GHz	180 GHz	40 GHz
Max. frequency of oscillation	f_{max}	250 GHz	250 GHz	120 GHz
Base-collector capacitance	C_{BC}	5.8 fF	5.2 fF	3.5 fF
Breakdown voltage	BV_{CE0}	1.6 V	1.7 V	4.0 V
Breakdown voltage	BV_{CB0}	5.8 V	6.5 V	15 V

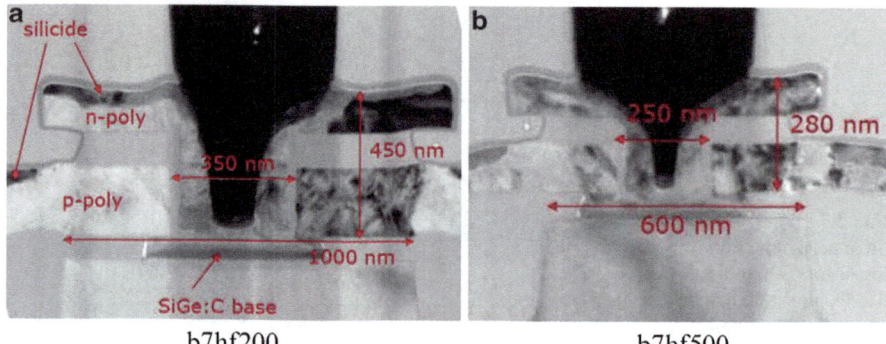

b7hf200   b7hf500

Fig. 3.9 Cross section of the emitter-base complex for an npn HBT. (**a**) b7hf200 with 200 GHz f_t / 250 GHz f_{max}. (**b**) b7hf500 process with 220 GHz f_t / 285 GHz f_{max}

The technology uses a double-polysilicon self-aligning concept [11]. Base and emitter are contacted by p^+ and n^+ polysilicon, respectively, to reduce the base resistance and parasitic capacitances. The actual emitter size is shrunk below the minimum feature width through the use of composite spacers. This increases the high-frequency performance of the transistors. Deep and shallow trench isolation is implemented to substantially decrease the collector-substrate capacitance. For the emitter, base, and collector the doping materials arsenide, boron, and phosphor are used, respectively. Furthermore, Ge is integrated into the base to form the HBT and enable bandgap engineering as discussed above. A small concentration of carbon is incorporated into the base in order to ensure a steep cutoff of the base doping profile during the following process steps. The use of two epitaxial layers instead of a single one enables simultaneous integration of high-speed and high-voltage transistors [12]. This second layer helps to decouple the adjustment of the thickness of high-speed and high-voltage transistors which results in an increased freedom regarding the tradeoff between transit frequency and breakdown voltage. Table 3.1 shows a comparison of the individual npn transistor types.

The b7hf500 technology is based on the b7hf200 process, but uses a smaller lithographic node. Figure 3.9 shows a comparison between the two transmission electron microscopy (TEM) cross sections for the npn transistor emitter-base complex.

A 250 nm minimum feature size in the b7hf500 technology is used for vertical scaling. This improves the aspect ratio and enables lateral scaling which reduces the base-collector and base-emitter capacitances c_{BC} and c_{BE} as well as the base

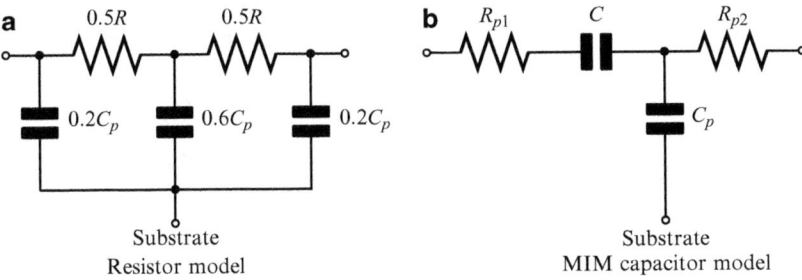

Fig. 3.10 Distributed simulation models in b7hf200. (**a**) Resistor. (**b**) MIM capacitor

resistance r_B. The transit frequency f_t is improved by increasing the Ge content in the base and a reduced thermal budget which limits dopant diffusion. Additionally, the width of the emitter-base spacer is shrunk to further lower the base resistance.

3.2.2 Passive Elements

Two different polysilicon resistors and a high accuracy thin film TaN resistor are provided by the technology. The poly resistors are processed as base polysilicon with doping levels p^+ or p^-. They posses an area specific capacitance of $150\Omega/\square$ and $1,000\Omega/\square$, respectively, with manufacturing tolerances up to $\pm20\%$. In contrast, the TaN resistor shows an area specific resistance of $20\Omega/\square$ with tolerances of $\pm8\%$. Figure 3.10a shows the implemented resistor model. In this distributed model R denotes the total resistance while C_p represents the total parasitic capacitance to the substrate.

Metal-insulator-metal (MIM) capacitors are available in the b7hf200 process which are located between the second and third metal layer. The insulator is formed by a 50 nm thin dielectric material that consists of Al_2O_3. Figure 3.10b depicts the utilized model of the MIM capacitor. The capacitance of the MIM structure is denoted as C while the resistors R_{p1} and R_{p2} represent the parasitic contact resistances. Their values depend on the physical dimension and configuration of the capacitor. The close proximity of the second metal layer to the substrate is modeled by the parasitic capacitance C_p which is about 2% of the MIM capacitance C. In addition the process provides a varactor with high quality factor for oscillator circuits.

The b7hf200 process features five metallization layers. The top layer is realized as aluminum and is used to form the pads and to implement laser fuses. The remaining metal layers are made of copper and used as interconnect. The topmost copper layer exhibits the maximum thickness and therefore highest current carrying capability.

Microstrip lines can be implemented into the layer stack to realize high-frequency signal interconnections and inductive elements. Figure 3.11a shows a microstrip line configuration as used throughout this work. The signal line is usually

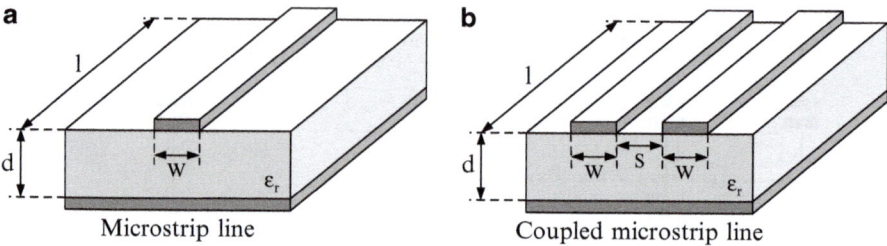

Fig. 3.11 Configuration of a microstrip line. The characteristic impedance is altered by changing width W or distance d to the ground plane. (**a**) Standard. (**b**) Coupled microstrip line

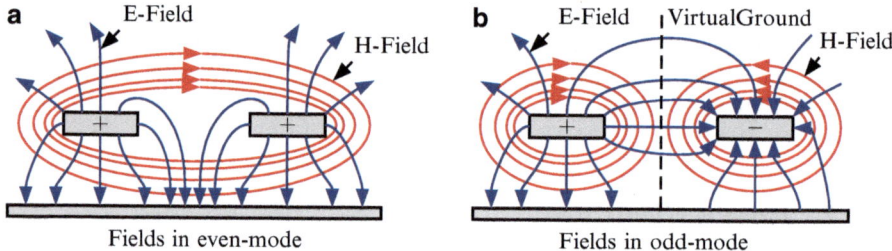

Fig. 3.12 Distribution of the electric field E and magnetic field H in a differential microstrip line cross section. (**a**) Even-mode. (**b**) Odd-mode

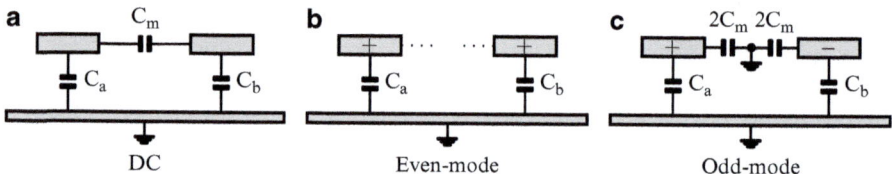

Fig. 3.13 Capacitances per unit length of a differential microstrip line in the (**a**) DC case. (**b**) Even-mode. (**c**) Odd-mode

formed by the topmost copper layer CU4 and separated from the CU2 ground plane by the dielectric material SiO$_2$ with a relative permittivity of $\varepsilon_r = 3.9$. The characteristic impedance of the transmission line can be altered by adjusting the width W of the line or by changing the distance d to the ground plane using another metal layer.

Coupled (differential) microstrip lines consist of two parallel lines above a common ground plane as shown in Fig. 3.11b. They are used for differential routing and the coupling of signals. The characteristic impedance Z_0 of differential microstrip lines is described by a model of odd (both signals are equal in amplitude and phase) and even (both signals are 180° out of phase) impedances Z_o and Z_e. These modes differ in the field distribution between the lines as depicted in Fig. 3.12.

Therefore, even- and odd-mode signals experience different capacitances per unit length and consequently a different characteristic impedance as shown in Fig. 3.13. A virtual ground is present at the intersecting line for odd-mode operation.

References

1. M. Reisch, *High-Frequency Bipolar Transistors: Physics, Modeling, Applications.* Springer-Verlag, 2003.
2. J. J. Ebers and J. L. Moll, "Large-signal behavior of junction transistors," *Proc. IRE*, vol. 42, no. 12, pp. 1761–1772, Dec. 1954.
3. U. Tietze and C. Schenk, *Halbleiter-Schaltungstechnik, (in German)*, 13rd ed. Springer-Verlag, 2010.
4. H. Gummel and H. Poon, "A compact bipolar transistor model," in *IEEE Int. Solid-State Circuits Conf. Dig. Tech. Papers*, San Francisco, CA, Feb. 1970, pp. 78–79.
5. P. Ashburn, *SiGe Heterojunction Bipolar Transistors.* John Wiley & Sons, 2003.
6. B. Geynet, P. Chevalier, B. Vandelle, F. Brossard, N. Zerounian, M. Buczko, D. Gloria, F. Aniel, G. Dambrine, F. Danneville, D. Dutartre, and A. Chantre, "SiGe HBTs featuring $f_T > 400\,GHz$ at room temperature," in *Proc. Bipolar/BiCMOS Circuits Technol. Meeting*, Monterey, CA, Oct. 2008, pp. 121–124.
7. J. D. Cressler, *Silicon Heterostructure Handbook: Materials, Fabrication, Devices, Circuits, and Applications of SiGe and Si Strained-Layer Epitaxy.* CRC Press, 2006.
8. ——, "SiGe HBT technology: A new contender for Si-based RF and microwave circuit applications," *IEEE Trans. Microw. Theory Tech.*, vol. 46, no. 5, pp. 572–589, May 1998.
9. J. Böck, H. Schäfer, K. Aufinger, R. Stengl, S. Boguth, R. Schreiter, M. Rest, H. Knapp, M. Wurzer, W. Perndl, T. Böttner, and T. F. Meister, "SiGe bipolar technology for automotive radar applications," in *Proc. Bipolar/BiCMOS Circuits Technol. Meeting*, Montreal, Canada, Sep. 2004, pp. 84–87.
10. M. Hartmann, "Analysis and design of monolithic integrated SiGe mixer circuits for 77 GHz automotive radar," Ph.D. dissertation, Inst. for Electron. Eng., Univ. of Erlangen-Nuremberg, Erlangen, Germany, 2007.
11. J. Böck, H. Schäfer, H. Knapp, K. Aufinger, M. Wurzer, S. Boguth, T. Böttner, R. Stengl, W. Perndl, and T. F. Meister, "3.3 ps SiGe bipolar technology," in *IEEE Int. Electron Devices Meeting Tech. Dig.*, San Francisco, CA, Dec. 2004, pp. 255–258.
12. R. K. Vytla, T. F. Meister, K. Aufinger, D. Lukashevich, S. Boguth, J. Böck, H. Schäfer, and R. Lachner, "Simultaneous integration of SiGe high speed transistors and high voltage transistors," in *Proc. Bipolar/BiCMOS Circuits Technol. Meeting*, Maastricht, The Netherlands, Oct. 2006.

Chapter 4
Millimeter-Wave Receiver Concepts

4.1 Mixer Principle

Mixers are a central part of radio transmission systems. They are used for frequency translation in the transmit and receive path. Figure 4.1 shows the symbol of an ideal mixer that operates as a multiplier using a local oscillator. Frequency conversion of an input signal which is lower than the output signal is called up-conversion whereas the opposite case in which the output signal frequency is lower than the input frequency is known as down-conversion. Up-conversion mixers are usually implemented in the transmit path and are often referred to as modulators. On the other hand, down-conversion is performed in the receiver path to convert the received signal back into the baseband (demodulation). The lower frequency is denoted as intermediate frequency f_{IF} and the higher frequency is called radio frequency f_{RF}.

Mixers can be separated into additive and multiplying configurations via their transfer function characteristics. In the following, only multiplying mixers will be analyzed, as they offer a number of advantages and form the basis of the majority of today's frequency converters. The mathematical principle of these mixers lies in the multiplication of two periodic signals. The input signal is multiplied with the local oscillator signal s_{LO}, with the latter usually providing a constant power level. Thereby, two signals located at the sum and the difference of both frequencies are generated according to (4.1) for products of trigonometric functions.

$$\cos\alpha\cos\beta = \frac{1}{2}\left(\cos(\alpha-\beta)+\cos(\alpha+\beta)\right) \tag{4.1}$$

For up-conversion the IF signal

$$s_{IF}(t) = a(t)\cos\left(\omega_{IF}t+\varphi(t)\right) \qquad \text{with} \quad \omega_{IF} = 2\pi f_{IF} \tag{4.2}$$

D. Kissinger, *Millimeter-Wave Receiver Concepts for 77 GHz Automotive Radar in Silicon-Germanium Technology*, SpringerBriefs in Electrical and Computer Engineering, DOI 10.1007/978-1-4614-2290-7_4, © Springer Science+Business Media, LLC 2012

Fig. 4.1 Graphic symbol of a mixer with signal denotation; up-conversion mixer: $s_{in} = s_{IF}$, $s_{out} = s_{RF}$; down-conversion mixer: $s_{in} = s_{RF}$, $s_{out} = s_{IF}$

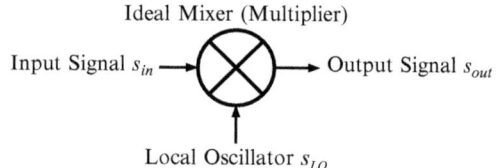

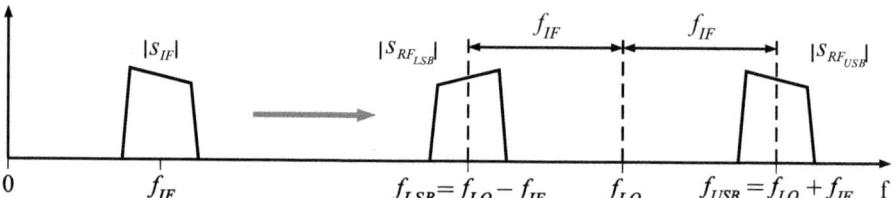

Fig. 4.2 Spectral distribution of the signals present in an up-conversion mixer

at the input is mixed with the LO signal

$$s_{LO}(t) = 2\cos(\omega_{LO}t) \qquad \text{with} \quad \omega_{LO} = 2\pi f_{LO}. \tag{4.3}$$

Therefore at the output the following RF signals are present:

$$s_{RF}(t) = s_{IF}(t)s_{LO}(t) = a(t)\cos\big(\omega_{IF}t + \varphi(t)\big)2\cos(\omega_{LO}t) =$$

$$= \underbrace{a(t)\cos\big((\omega_{LO}+\omega_{IF})t + \varphi(t)\big)}_{uppersideband\,(f>f_{LO})} + \underbrace{a(t)\cos\big((\omega_{LO}-\omega_{IF})t - \varphi(t)\big)}_{lowersideband\,(f<f_{LO})}. \tag{4.4}$$

In the upper sideband (USB) one yields a signal at the sum of both frequencies $f_{USB} = f_{LO} + f_{IF}$. The signal in the lower sideband (LSB) lies at difference of the frequencies $f_{LSB} = f_{LO} - f_{IF}$. Either sideband carries the information of the IF signal. For transmission one of the two sidebands is suppressed. Figure 4.2 shows the spectral distribution of the signals present in an up-converter.

During down-conversion the RF signal at the input:

$$s_{RF}(t) = a(t)\cos\big(\omega_{RF}t + \varphi(t)\big) \qquad \text{with} \quad \omega_{RF} = 2\pi f_{RF}, \tag{4.5}$$

is multiplied with the LO signal

$$s_{LO}(t) = 2\cos(\omega_{LO}t) \qquad \text{with} \quad \omega_{LO} = 2\pi f_{LO}. \tag{4.6}$$

Figure 4.3 shows the spectral distribution of the signals present in a down-conversion mixer. For upper sideband transmission ($f_{RF} > f_{LO}$) one yields:

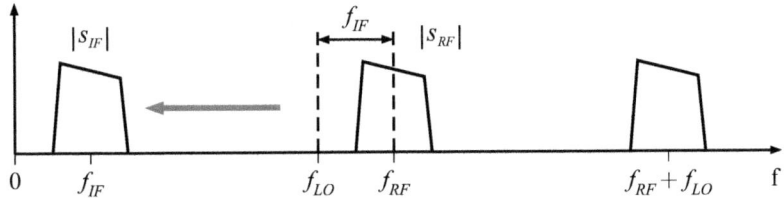

Fig. 4.3 Spectral distribution of the signals present in a down-conversion mixer

$$s_{IF}(t) = s_{RF}(t)s_{LO}(t) = a(t)\cos\left(\omega_{RF}t + \varphi(t)\right)2\cos(\omega_{LO}t)$$

$$= a(t)\cos\left((\omega_{RF} + \omega_{LO})t + \varphi(t)\right) + a(t)\cos\left((\omega_{RF} - \omega_{LO})t + \varphi(t)\right). \quad (4.7)$$

The additional output signal at the sum of both frequencies has to be suppressed by a subsequent low-pass filter. After ideal filtering the output signals becomes:

$$s_{IF}(t) = a(t)\cos\left((\omega_{RF} - \omega_{LO})t + \varphi(t)\right). \quad (4.8)$$

4.2   Receiver Topologies

Receivers (RX) represent the input path of a radiocommunication system. The main task of a receiver is filtering, down-conversion, and demodulation of the desired signal from a sum of interferers and noise. Demodulation refers to the recovery of the wanted signal information from the transmitted carrier. In modern transceivers this task as well as subsequent signal processing is performed in the digital domain. Because of the strong attenuation of the originally transmitted signal through the radio channel, amplification of the received signal has to be provided to meet the appropriate signal levels for digital post-processing of the data. Figure 4.4 shows a block diagram of a receiver architecture.

In most cases the RF signal received by the antenna has a relatively low power level. The signal is therefore amplified by a low-noise amplifier (LNA). Its main task is amplification of the RF signal with sufficiently low noise figure *NF* for suppression of the high noise contributions of the subsequent mixer stage. By means of a local oscillator (LO) the amplified signal is translated into the lower baseband frequency by the mixer. The down-converted signal can be digitized by an analog-to-digital converter (ADC) and post-processed in a digital signal processor (DSP).

Different techniques for conversion of the received RF signal into the baseband exist [1]. In the following sections two selected topologies for receivers will be presented, namely superheterodyne receivers and direct-conversion architectures.

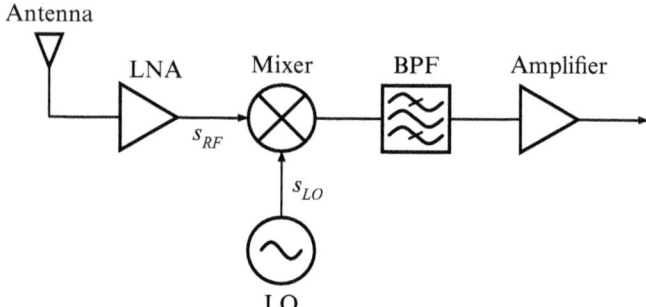

Fig. 4.4 General block diagram of a receiver consisting of LNA: low-noise amplifier, mixer, LO: local oscillator, BPF: bandpass filter, and IF amplifier

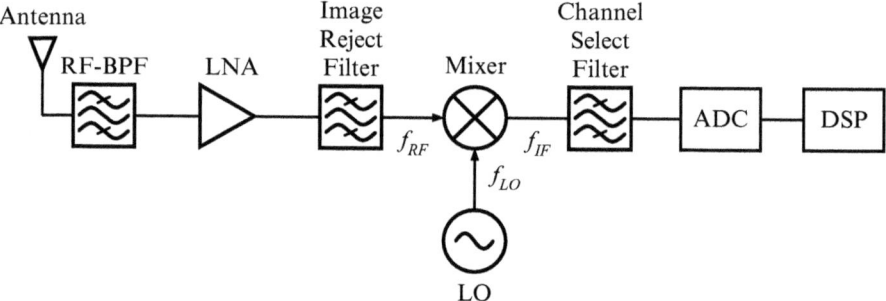

Fig. 4.5 Simplified block diagram of a superheterodyne receiver. The second frequency translation takes place in the digital domain via the signal processing unit (DSP). LNA: *Low-Noise Amplifier*, LO: *Local Oscillator*, ADC: *A/D Converter*, DSP: *Digital Signal Processor*

4.2.1 Superheterodyne Receiver

Superheterodyne receivers first convert the detected signal onto a constant intermediate frequency (IF) by means of a LO frequency which is different to the carrier frequency f_c. In a second conversion step the IF signal is further down-converted into the baseband. The main advantage of superheterodyne receivers is a simple mirror frequency suppression, as well as a constant frequency difference between the IF frequency and the baseband. A majority of today's receivers are based on the above principle. Figure 4.5 shows a block diagram of a superheterodyne receiver.

Unwanted out-of-band spectral content received through the antenna is filtered out by the RF bandpass filter (RF-BPF) to relieve the subsequent low-noise amplifier from undesired strong interferers which cause intermodulations and can drive the LNA into saturation. The filter can be implemented as an antenna with a small bandwidth or a dedicated SAW device. After the signal has been amplified by the low-noise amplifier an additional bandpass filter (*Image Reject Filter*) suppresses the mirror frequency. The resulting signal s_{RF} is mixed with the local oscillator (LO)

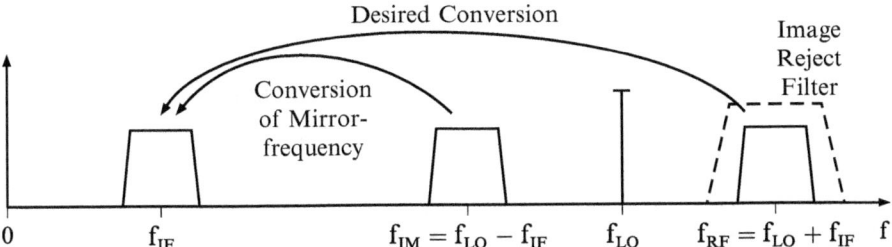

Fig. 4.6 Spectral distribution of the specific signals in a superheterodyne receiver

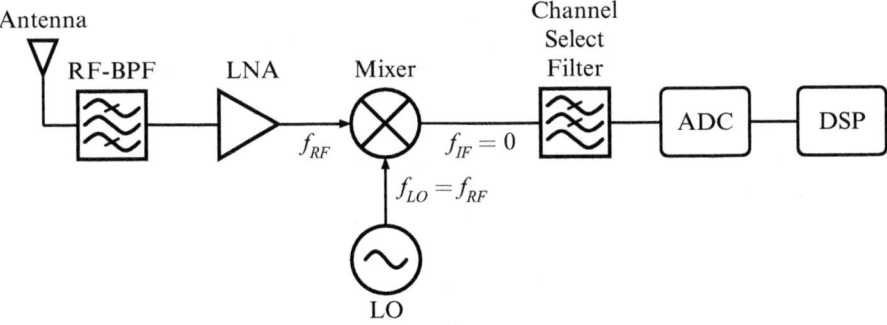

Fig. 4.7 Block diagram of a direct-conversion receiver architecture

signal s_{LO}. Simultaneously the frequency of the LO signal differs from the RF signal s_{RF} by a frequency offset of f_{IF}. Therefore a signal s_{IF} at the intermediate frequency f_{IF} is present at the output of the mixer stage. Via the *Channel Select Filter* the frequency range of the desired channel can be selected. The digitized signal at the output of the ADC is processed in the DSP to acquire the information located in the baseband. Figure 4.6 shows the working principle of a superheterodyne receiver with the help of the spectral distribution of the specific signals. The image reject filter avoids unwanted conversion of spectral content at the image frequency f_{IM} to be converted into the intermediate frequency band f_{IF}.

4.2.2   Direct-Conversion Receiver

Homodyne receivers are also known as direct-conversion receivers. The name is due to its working principle. Direct-conversion receivers operate at an intermediate frequency $f_{IF} = 0$. Hence they are also known as *zero-IF receiver*. Figure 4.7 shows a block diagram of a direct-conversion receiver. The carrier frequency f_c coincides with the LO frequency, therefore the received RF signal is converted directly into the baseband. The channel select filter can be implemented as a low-pass filter. The spectral distribution of the signals is depicted in Fig. 4.8.

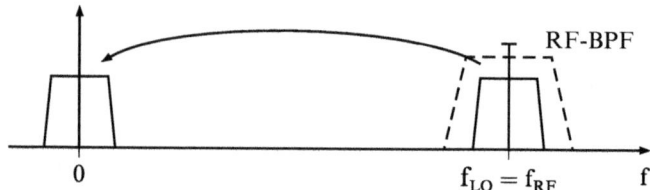

Fig. 4.8 Spectral distribution of the specific signals in a direct-conversion receiver

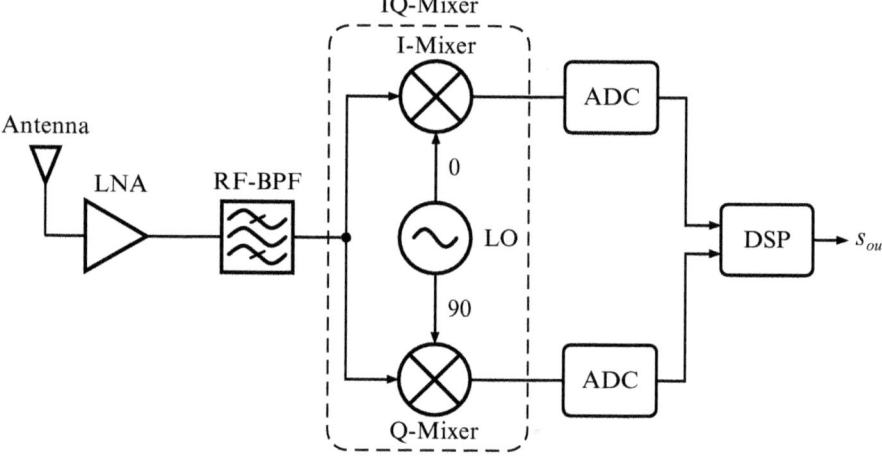

Fig. 4.9 Block diagram of an IQ-receiver for direct quadrature down-conversion

The simple homodyne architecture offers two important advantages over heterodyne architectures. The problem of image frequencies is circumvented because $f_{IF} = 0$, therefore no image filter is required. Second, baseband filtering can be performed by a low-pass filter and subsequent IF amplifiers which are easy to implement with integrated technologies. However, direct-conversion receivers suffer from some important drawbacks. Due to the fact that $f_{LO} = f_{RF}$ leakage from the local oscillator into the RF path results in unwanted radiation over the antenna as well as in self-mixing which causes DC offsets at the output of the mixing stage.

The circuit overlaps positive and negative parts of the input spectrum which causes loss of information for modulation schemes other than double-sideband AM modulation. For frequency and phase-modulation signals the down-conversion mixer has to provide quadrature phases because the two different sides of the spectrum carry different information and have to be separated accordingly. Figure 4.9 shows a block diagram of a direct-conversion IQ-receiver capable of separation of the two-sided spectrum. The LO signal is shifted by a phase of 90° to generate in-phase and quadrature signals at the output of the receiver. Errors in the phase relationship as well as amplitude mismatch (I/Q imbalances) corrupt the signal.

4.3 Receiver Performance Parameters

Mixers are characterized through a number of different parameters. The most important performance metrics are described in this section. Many of the parameters below are also relevant for other receiver building blocks, e.g. amplifiers.

4.3.1 Conversion Gain

The gain G of a two-port is generally defined as the ratio of output to input power. Although a mixer is essentially a three-port device, it can be regarded as a frequency translating two-port for a constant power level at the LO port. It is usually given as

$$G = 10\log_{10}\frac{P_{out}}{P_{in}}\,\text{dB}\,. \qquad (4.9)$$

In high-frequency circuits mismatches between the ports and the source/load lead to a partial reflection of energy at transitions of the characteristic impedance. Thus only a part of the source power can be transferred to the load, leading to different definitions of gain in high-frequency devices [2]. The transfer behavior of amplifiers is described by the transducer power gain G_T which is defined as follows:

$$G_T = \frac{P_L}{P_{A,g}} = \frac{\text{power available from the two-port network}}{\text{power available from the source}}\,. \qquad (4.10)$$

The conversion gain G_C describes the ratio of output to input signal power of frequency converting devices, e.g. mixers. The LO input is not considered in this description, although the conversion gain strongly depends on the LO power level:

$$G_C = 10\log_{10}\frac{P_{out}}{P_{in}}\,. \qquad (4.11)$$

The above equation can be applied to the ratio of voltage levels by application of the ohmic law, yielding the following definition:

$$G_C = 10\log_{10}\frac{v_{out}^2 R_{in}}{v_{in}^2 R_{out}}\,. \qquad (4.12)$$

For the case that the input resistance R_{in} is equal to the output resistance R_{out}, (4.12) can be simplified as follows:

$$G_{VC} = 20\log_{10}\frac{v_{out}}{v_{in}}\,. \qquad (4.13)$$

The above definition is referred to as voltage conversion gain G_{VC}. Note that it is only equal to the power-related conversion gain G_C if input and output impedance of the receiver are of the same value, an assumption that does not always hold true.

4.3.2 Noise Figure

The noise figure of a two-port is defined as the ratio of the signal-to-noise ratio (SNR) of the input related to the output of the device. It is a measure for the degradation of signal-to-noise ratio in a system. Two definitions are used for description of the above. The *Noise Factor F* represents the linear relationship

$$F = \frac{SNR_{in}}{SNR_{out}} = \frac{S_{in}N_{out}}{S_{out}N_{in}}, \tag{4.14}$$

whereas the *Noise Figure NF* denotes the logarithmic ratio in dB:

$$NF = 10\log_{10}\frac{SNR_{in}}{SNR_{out}}. \tag{4.15}$$

In addition to the above, mixers have two different definitions of noise figure. A distinction is drawn between single sideband noise figure NF_{SSB} and double sideband noise figure NF_{DSB}. The single sideband noise figure is given if the signal power of the desired signal is only present in one of the two sidebands. In this case the signal power from the single band is given relative to the total noise power present in both sidebands. Similarly the double sideband noise figure is used if the desired signal power is located in both sidebands. Either definition is related to the same noise power, with the distinction that the SSB signal power is only half as high, yielding:

$$NF_{SSB} = NF_{DSB} + 3\,\text{dB}. \tag{4.16}$$

As mentioned previously receiver gain and noise figure are strongly dependent on the LO power. Figure 4.10 shows a typical relation between these parameters.

4.3.3 Linearity and Intermodulation

The transfer behavior of many high-frequency transceiver components can be approximated with a linear model to obtain their small-signal response. Nevertheless nonlinearities of these devices often lead to important phenomenas [3]. Such systems can generate frequency components that do not exist in the input signal (fundamental). If a sinusoid is applied to a nonlinear system ($x = A\cos\omega t$),

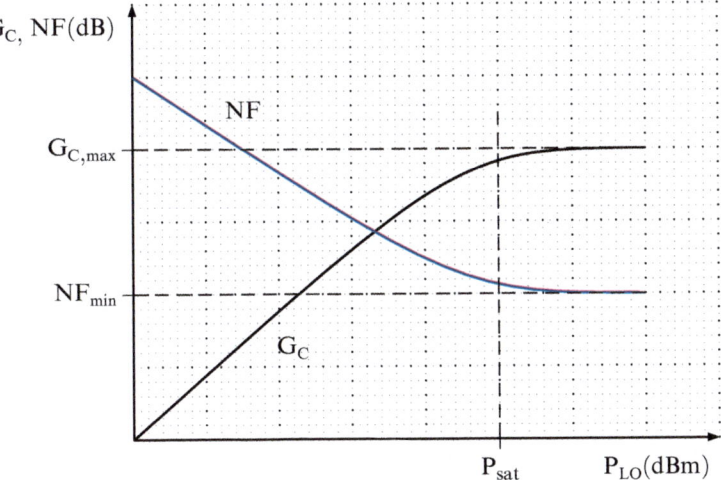

Fig. 4.10 Typical dependency of receiver conversion gain G_C and noise figure NF as a function of the applied local oscillator (LO) power

Fig. 4.11 Diagram of a signal at the output of an amplifier which is subject to clipping

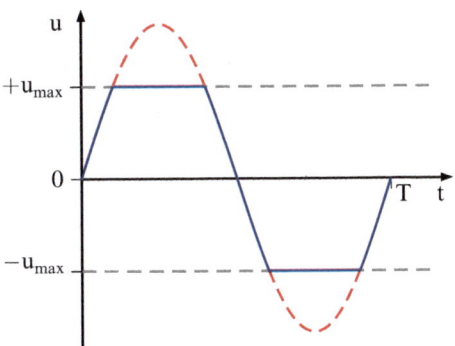

its output exhibits frequency components that are integer multiples of the input frequency:

$$y(t) = \alpha_1 A \cos \omega t + \alpha_2 A^2 \cos^2 \omega t + \alpha_3 A^3 \cos^3 \omega t + \cdots$$

$$= \alpha_1 A \cos \omega t + \frac{\alpha_2 A^2}{2}(1 + \cos 2\omega t) + \frac{\alpha_3 A^3}{4}(3 \cos \omega t + \cos 3\omega t) \qquad (4.17)$$

$$= \frac{\alpha_2 A^2}{2} + \left(\alpha_1 A + \frac{3\alpha_3 A^3}{4}\right)\cos \omega t + \frac{\alpha_2 A^2}{2}\cos 2\omega t + \frac{\alpha_3 A^3}{4}\cos 3\omega t. \quad (4.18)$$

The first term in (4.18) is the DC value of the output. The second term at the input frequency is the fundamental part. The higher order terms are called harmonics. They arise for example through overdrive of an amplifying stage. Figure 4.11 shows a time-domain signal form at the output of an amplifier driven into saturation. A maximum voltage level u_{max} causes the amplifier to clip parts of the sine wave.

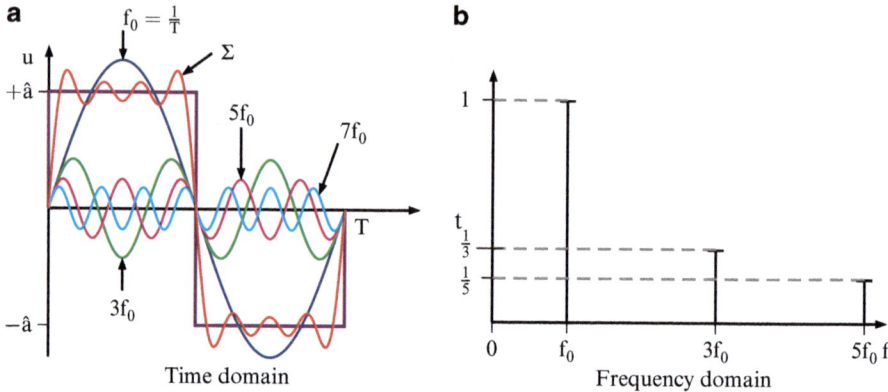

Fig. 4.12 Representation of a square wave as the sum of harmonics. (**a**) Time domain. (**b**) Frequency domain, limited to the fifth harmonic

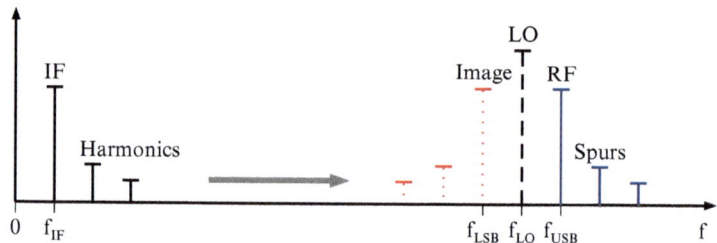

Fig. 4.13 Typical spectrum of an up-conversion mixer (modulator) that is subject to harmonics at the IF input and subsequently spurs at the RF output

With increasing overdrive the signal approximates a square wave form. Decomposition of the square wave into a Fourier series yields the following mathematical expression for the spectrum of the harmonic distortions:

$$f(t) = \frac{4\hat{a}}{\pi} \sum_{n=1}^{\infty} \frac{\sin((2n-1)\omega t)}{2n-1} \qquad \text{with} \quad \omega = 2\pi f$$

$$= \frac{4\hat{a}}{\pi} \left[\sin \omega t + \frac{1}{3} \sin 3\omega t + \frac{1}{5} \sin 5\omega t + \cdots \right]. \tag{4.19}$$

The representation of a square wave signal as the sum of sine waves in the time and frequency domain is depicted in Fig. 4.12. In this example the resulting waveform for a summation of harmonics up to the seventh order is shown.

Mixed harmonics of the fundamental input frequency lead to undesired signals in the output spectrum of a mixer. These signals are called spurs. Figure 4.13 shows a typical spectrum of an up-conversion mixer with an input signal that contains harmonics. The additional harmonic content is up-converted into the upper and lower sideband and appears as unwanted spurious emissions.

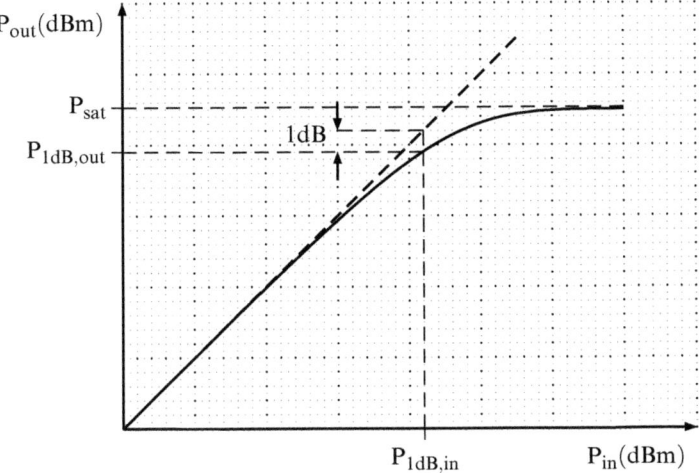

Fig. 4.14 Diagram of the typical large-signal behavior of a device with a 1 dB input-referred compression point $P_{1dB,in}$ and a maximum saturated output power P_{sat}

The input- or output-referred compression point of a receiver represents one means to describe its nonlinear behavior. By specification of the compression point of a mixer, one can define the area in which the mixer operates in a linear fashion. Usually the input-referred 1-dB compression point is given. It represents the applied input power that leads to a 1 dB degradation in the receiver power gain. The above phenomena is schematically depicted in Fig. 4.14. For small input power levels the transfer function of receiver can be regarded as quasi-linear. With an increase in input power the gain of the device is compressed until it approaches saturation.

Besides the desired mixing products of the fundamental input frequencies and the LO signal, the output of the mixer exhibits components that are not harmonics of the input frequencies. If two input signals with different frequencies are applied to a nonlinear system intermodulation distortion occurs. Resulting from multiplication of the two signals, these additional components occur at the following frequencies:

$$f_{m+n} = nf_1 \pm mf_2 \quad \text{with} \quad n, m \in \mathbb{N}. \quad (4.20)$$

The sum $m + n$ indicates the order of the intermodulation product. In a mixer operated in the nonlinear region these intermodulation products are down-converted into the IF. Figure 4.15 shows the spectrum at the input and output of the mixer.

Out of the number of the additional components, the third order intermodulation products (IM3) located at $2f_{IF_1} - f_{IF_2}$ and $2f_{IF_2} - f_{IF_1}$ are usually the most relevant, as they occur close to the desired signals and exhibit a relatively high power level.

Figure 4.16 shows the output powers of the desired signal and the third order intermodulation product (IM3) versus the fundamental input power. With the slope of the third order intermodulation product being steeper than that of the desired

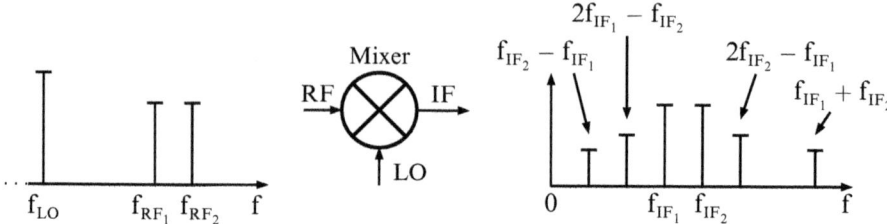

Fig. 4.15 Intermodulation in a mixer subject to a two-tone test

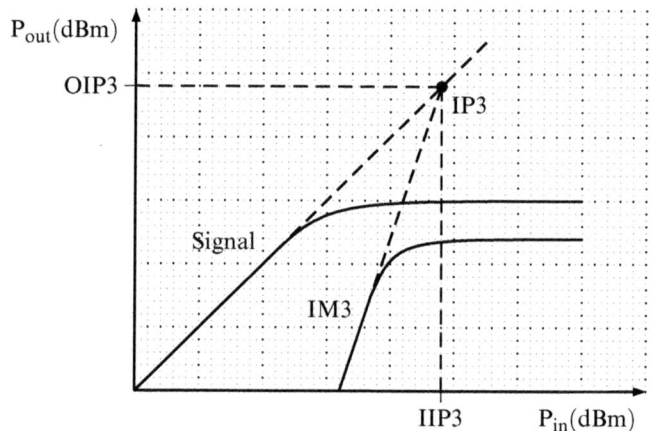

Fig. 4.16 Output power of desired signal and third order intermodulation product (IM3) versus input power. The extrapolated intercept point IP3 is depicted graphically

output signal, a virtual point can be defined through extrapolation, where the output power of the intermodulation product approaches that of the desired signal. This point is referred to as the third order input intercept point (IIP3) or output intercept point (OIP3), respectively. For a majority of receivers the IM3 is the most relevant, but similar observations can be made for higher order intermodulation products.

4.3.4 Port Matching and Isolation

Another important property is impedance matching at the high-frequency ports of the receiver. To obtain optimum power transfer from the source or load to the two-port network one has to ensure impedance matching at the corresponding ports. Mismatches result in a partial reflection of the available power which causes a reduction in the overall gain of the system. The RF input port of a receiver is especially vulnerable to mismatches as an increased noise figure due to additional degradation of the signal-to-noise ratio results. Mismatches at the LO input can usually be tolerated to a certain amount and compensated for by providing an increased input power.

Port-to-port isolation of a receiver describes the decoupling of two selected signal paths with respect to each other. Limited isolation in receivers leads to a number of phenomenas that have to be considered. In a direct-conversion receiver architecture coupling from the LO to the RF port, called LO leakage, results in unwanted radiation of the LO signal through the antenna as well as DC offsets due to self-mixing of the oscillator signal. Parts of the RF input signal may also directly couple to the output port, leading to additional spectral content at the intermediate frequency.

4.4 Differential Architectures

A system has odd symmetry if its response to an input signal $-x(t)$ is the negative of that to $x(t)$. A device that provides the above transfer characteristics is called *differential* or *balanced* circuit and cancels out even order terms in (4.18). In a differential circuit signals with equal magnitude and opposite phase relative to the reference ground are processed [4]. Figure 4.17 shows a simple balanced circuit comprised of two single-ended signal sources with opposite phase and equal amplitude. Both load resistors are connected to a common ground. The ideal circuit does not exhibit any signal current flow through the ground references. Therefore, these nodes can also be of a virtual nature.

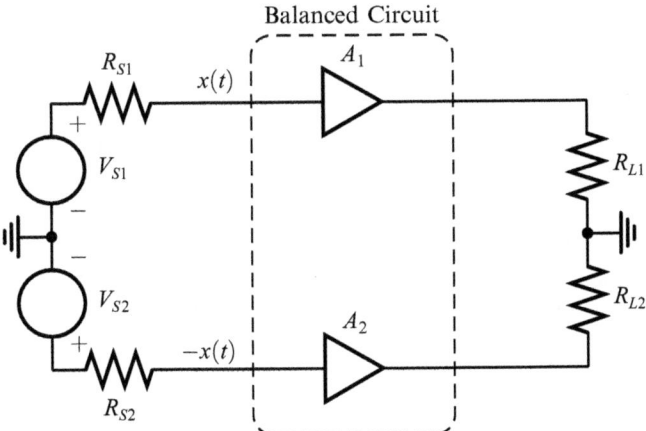

Fig. 4.17 Balanced circuit representation of an amplifier with differential source and load. The reference grounds can be of real or virtual nature

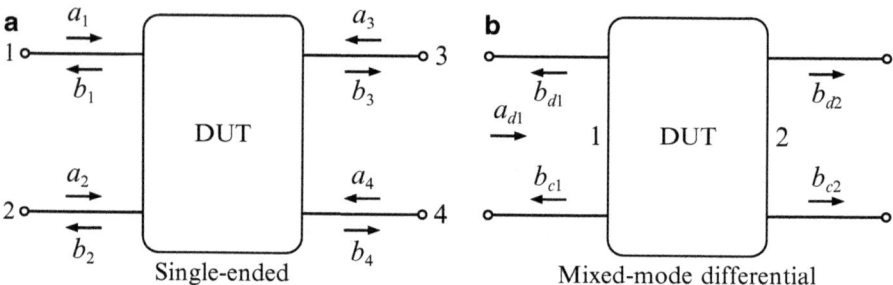

Fig. 4.18 Power wave representation. (**a**) Single-ended four-port circuit. (**b**) Differential two-port circuit with mixed-mode differential excitation at the input

Differential circuit techniques offer a number of advantages over single-ended designs. Among these are a 6 dB increase in distortion-free dynamic range for a given supply voltage, reduced even-order distortions, and superior noise cancellation properties. The latter can be characterized by the common-mode rejection ratio. Main drawbacks are the doubling of the circuitry and the power consumption.

Figure 4.18a shows the incident and reflected power waves at a single-ended four-port device. The relationship between incident and reflected waves a_i and b_j can be described by the scattering parameter matrix S^{std} in (4.21).

$$\begin{vmatrix} b_1 \\ b_2 \\ b_3 \\ b_4 \end{vmatrix} = \underbrace{\begin{vmatrix} S_{11} & S_{12} & S_{13} & S_{14} \\ S_{21} & S_{22} & S_{23} & S_{24} \\ S_{31} & S_{32} & S_{33} & S_{34} \\ S_{41} & S_{42} & S_{43} & S_{44} \end{vmatrix}}_{S^{std}} \begin{vmatrix} a_1 \\ a_2 \\ a_3 \\ a_4 \end{vmatrix} \tag{4.21}$$

In contrast to the former a differential circuit features one pair of ports at the input as well as one pair at the output, as depicted in Fig. 4.18b.

Mixed-mode S-parameters can illustrate and separate the differential, common-mode, and cross-mode signal relationships of a four-port device. The relationship between differential and common-mode power waves is described through the mixed-mode matrix S^{mm}, shown in (4.22). Purely differential mixed-mode parameters have the designator S_{ddij}, while S_{cdij} represents cross-mode S-parameters.

$$\begin{vmatrix} b_{d1} \\ b_{d2} \\ b_{c1} \\ b_{c2} \end{vmatrix} = \begin{vmatrix} S_{dd} & S_{dc} \\ S_{cd} & S_{dc} \end{vmatrix} \begin{vmatrix} a_{d1} \\ a_{d2} \\ a_{c1} \\ a_{c2} \end{vmatrix} = \underbrace{\begin{vmatrix} S_{dd11} & S_{dd12} & S_{dc11} & S_{dc12} \\ S_{dd21} & S_{dd22} & S_{dc21} & S_{dc22} \\ S_{cd11} & S_{cd12} & S_{cc11} & S_{cc12} \\ S_{cd21} & S_{cd22} & S_{cc21} & S_{cc22} \end{vmatrix}}_{S^{mm}} \begin{vmatrix} a_{d1} \\ a_{d2} \\ a_{c1} \\ a_{c2} \end{vmatrix} \tag{4.22}$$

The remaining ports have to be terminated appropriately in order to perform the mixed-mode measurements. A transformation between standard S-parameters S^{std} and mixed-mode parameters S^{mm} exists. Equation 4.23 gives the formalism to obtain mixed-mode parameters through single-ended four-port measurements.

$$S^{mm} = M S^{std} M^{-1} \quad \text{with} \quad M = \frac{1}{\sqrt{2}} \begin{vmatrix} 1 & -1 & 0 & 0 \\ 0 & 0 & 1 & -1 \\ 1 & 1 & 0 & 0 \\ 0 & 0 & 1 & 1 \end{vmatrix} \tag{4.23}$$

This algorithm enables the determination of differential scattering parameters with the use of single-ended measurement equipment. It is especially useful as true differential vector network analyzers (VNA) in the millimeter-wave frequency range are not readily available. Nevertheless one has to provide proper on-chip termination for the respective unused ports which necessitates some additional efforts.

References

1. B. Razavi, *RF Microelectronics*. Prentice Hall, 1998.
2. D. M. Pozar, *Microwave Engineering*, 3rd ed. John Wiley & Sons, 2004.
3. S. A. Maas, *Nonlinear Microwave and RF Circuits*, 2nd ed. Artech House, 2003.
4. W. R. Eisenstadt, B. Stengel, and B. M. Thompson, *Microwave Differential Circuit Design Using Mixed-Mode S-Parameters*, 3rd ed. Artech House, 2006.

Chapter 5
Differential 77-GHz High-Linearity Receiver Front-End

5.1 Introduction

Over the recent years several different architectures for SiGe-based down-conversion mixers have been published, resembling standard double-balanced Gilbert cells [1,2] as well as micromixer topologies with single-ended RF inputs [3,4]. Published receiver front-ends feature an additional low-noise amplifying stage prior to the mixer to reduce the overall noise figure of the receiver chain [5–10].

Hard specifications for the automotive environment define a large dynamic range of the received signal. Besides a low noise figure, this necessitates a high linearity of the receiver front-end. This chapter presents the design and measurement results of a high-linearity receiver front-end for 77 GHz FMCW radar systems [11].

5.2 Circuit Design

The proposed receiver consists of a cascade of an LNA followed by a mixer stage for direct down-conversion of the received signal as shown in Fig. 5.1. Subsequently the down-converted IF signal is low-pass filtered and amplified by the buffer.

Both the LNA and the mixing stage are designed differentially. The inputs for the RF and LO signals feature a $\lambda/2$ transmission line (not shown in the schematic) connected between the differential pads. This line acts as a balun and enables the circuit to be additionally driven in single-ended mode through the conversion of the $100\,\Omega$ differential impedance to $25\,\Omega$ for single-ended measurement equipment.

Figure 5.2a shows the schematic of the low-noise amplifier. It consists of a cascode stage composed of the differential transistors Q_1 in a common-emitter configuration and transistors Q_2 that form a common-base amplifier stage. Transistors Q_1 are inductively degenerated via transmission lines T_2 for simultaneous noise and power matching at the input of the LNA. The L-shape input matching network is realized through series capacitors C_1 and parallel transmission line T_1. In addition

D. Kissinger, *Millimeter-Wave Receiver Concepts for 77 GHz Automotive Radar in Silicon-Germanium Technology*, SpringerBriefs in Electrical and Computer Engineering, DOI 10.1007/978-1-4614-2290-7_5, © Springer Science+Business Media, LLC 2012

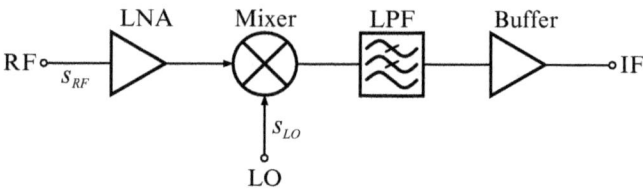

Fig. 5.1 General block diagram of the proposed receiver consisting of an *LNA* low-noise amplifier, mixer, *LO* local oscillator, *LPF* low-pass filter, and *IF* buffer

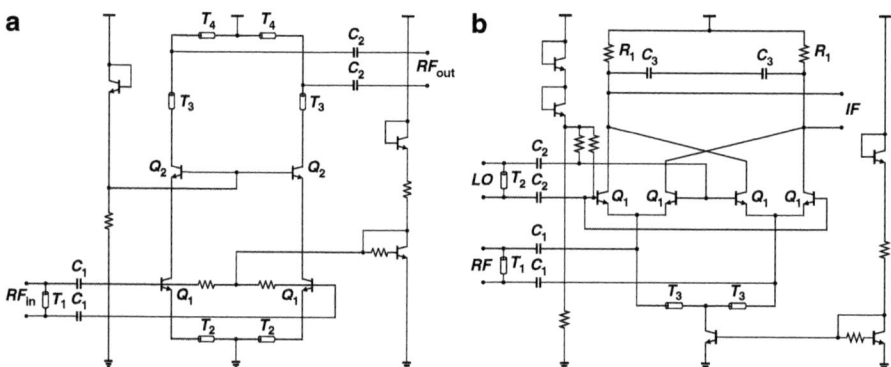

Fig. 5.2 (**a**) Schematic of the proposed differential cascode LNA. (**b**) Schematic of the proposed differential high-linearity double-balanced switching quad mixer

to the desired impedance transformation the capacitors C_1 enable integrated DC decoupling of the input port while the transmission line T_1 acts as an electrostatic discharge (ESD) protection at the input of the circuit. Capacitors C_2 in combination with transmission lines T_3 and T_4 form a T-type matching network at the output of the LNA. For the DC biasing of the common-emitter stage a direct biasing scheme with bipolar current mirrors has been used to avoid the loss of additional headroom necessary for a standard base point current source. The biasing for both cascode stages is fed through virtual ground nodes along the differential symmetry axis to avoid the necessity for additional AC ground capacitors.

The schematic of the implemented mixer is shown in Fig. 5.2b. It resembles a double balanced switching quad formed by transistors Q_1 with L-type matching networks C_1, T_1 for the RF and C_2, T_2 for the LO input port. The transconductance stage of a Gilbert mixer is omitted to enable high voltage swings at the mixer output at low supply voltages. Instead the RF signal path is decoupled from the current source through $\lambda/4$ lines T_3. The IF output is low-pass filtered through the combination of load resistors R_1 with capacitors C_3 and subsequently connected to a differential buffer composed of emitter followers (not shown) that transform the output impedance to $100\,\Omega$ (differential).

5.3   Experimental Results

The proposed circuits have been fabricated in a $200\,\mathrm{GHz}\ f_t\ /\ 250\,\mathrm{GHz}\ f_{max}$ automotive environment certified SiGe bipolar technology. Figure 5.3a, b show the layout and die photograph of the stand-alone differential cascode LNA.

The layout and die photograph of the designed stand-alone differential switching quad mixer is depicted in Fig. 5.4a, b. The overall chip size for both the LNA and the mixer is pad limited and consumes an area of $728 \times 728\,\mu\mathrm{m}^2$.

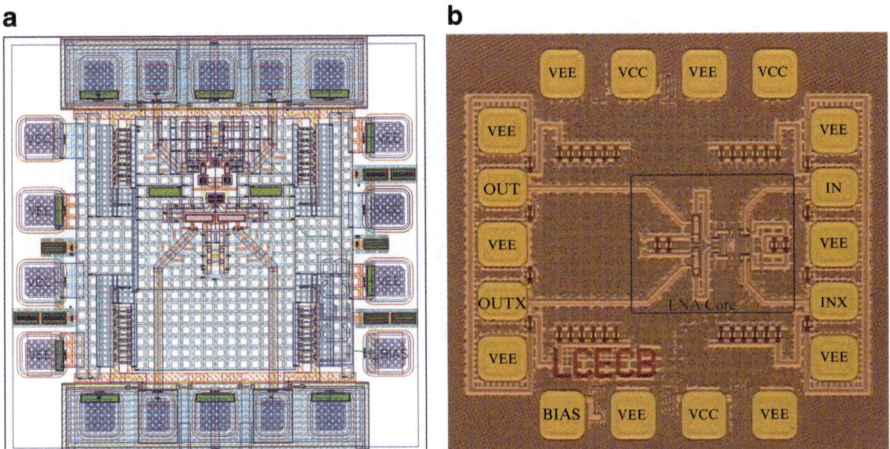

Fig. 5.3 Fabricated differential cascode LNA. (**a**) Layout. (**b**) Die photograph

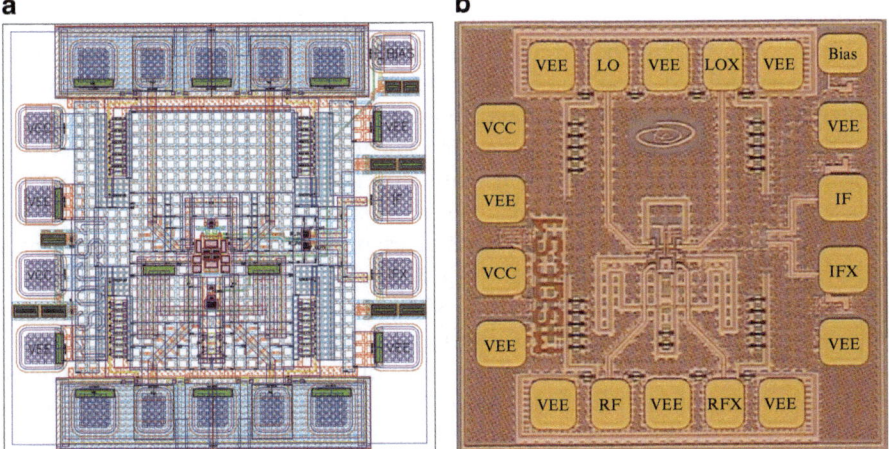

Fig. 5.4 Fabricated switching quad mixer. (**a**) Layout. (**b**) Die photograph

a   **b**

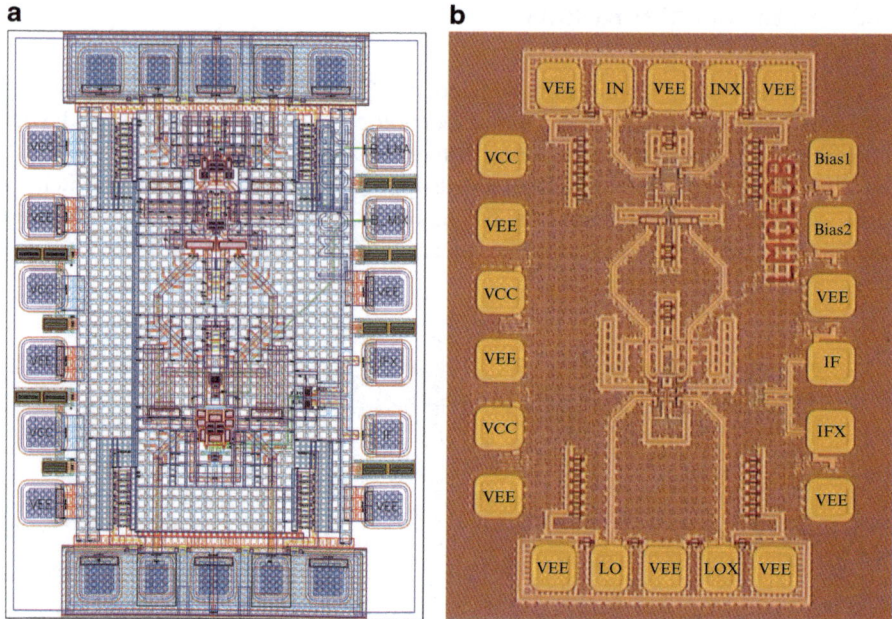

Fig. 5.5 Fabricated receiver front-end. (**a**) Layout. (**b**) Die photograph

Both LNA and mixer have been individually characterized for gain, noise, and linearity properties prior to the overall integration of the receiver front-end. An analysis regarding optimum bias points and matching bandwidth as well as investigations on different transistor sizes and matching elements have been conducted with the target of low-noise performance and simultaneous high-linearity properties.

Figure 5.5a, b show the layout and a die photograph of the receiver front-end. It has been realized as an integrated cascade of the above presented LNA and switching quad mixer. The overall pad limited chip area is $728 \times 1028\,\mu m^2$.

The characterization of the overall receiver front-end has been performed by single-ended on-wafer measurements with a measurement setup described in [12]. Noise and gain parameters have been measured at an IF of 4.8 MHz. A photograph of the receiver conversion gain and noise figure on-wafer measurement setup and a detailed close-up view are depicted in Fig. 5.6a, b.

Figure 5.7 shows the variation of the SSB noise figure and RF to IF conversion gain over the local oscillators power level for a fixed RF input level of $-16\,dBm$ at a frequency of 76.5 GHz. The mixing stage of the receiver shows no significant performance degradation for an LO power level above 0 dBm. At a fixed LO power of 0 dBm the SSB noise figure and conversion gain of the receiver front-end are 14 and 24 dB, respectively. Simulations show that the noise figure is further improved by 2 dB when the circuit is driven with a true differential RF source.

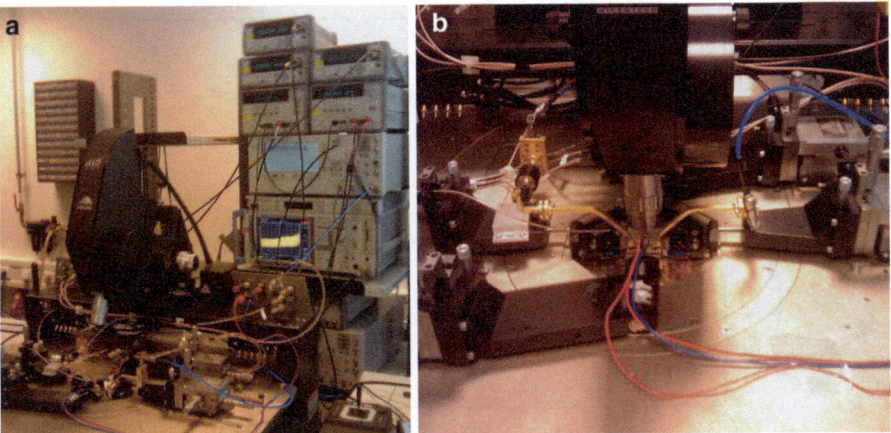

Fig. 5.6 Conversion gain and noise figure on-wafer characterization of the fabricated circuit. (**a**) Photograph of the measurement setup. (**b**) Detailed close-up view

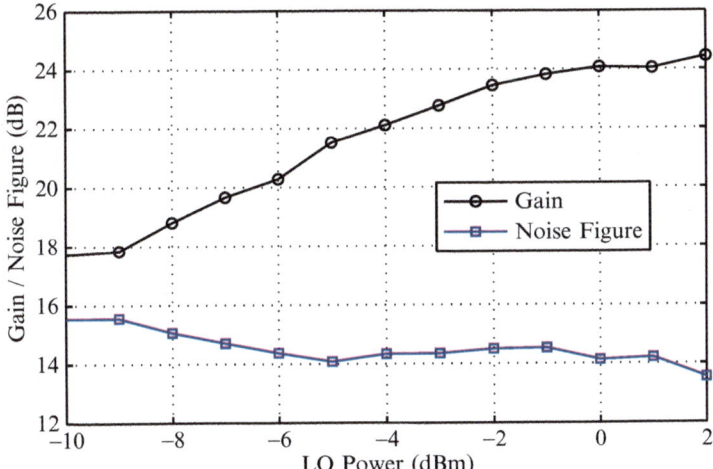

Fig. 5.7 Measurement results of conversion gain and SSB noise figure of the fabricated receiver front-end versus applied LO power at a frequency of 76.5 GHz. The RF input power has been fixed to a value of −16 dBm

Referring to Fig. 5.8 the front-end shows nearly constant behavior for the conversion gain and SSB noise figure across the intended frequency range from 76 to 77 GHz for a constant LO power of 0 dBm and −16 dBm RF input power.

Figure 5.9 shows the measured conversion gain of the fabricated receiver front-end versus the RF input power for an LO power of 0 dBm at the center frequency of 76.5 GHz. The input referred 1 dB compression point is shown to be −10 dBm.

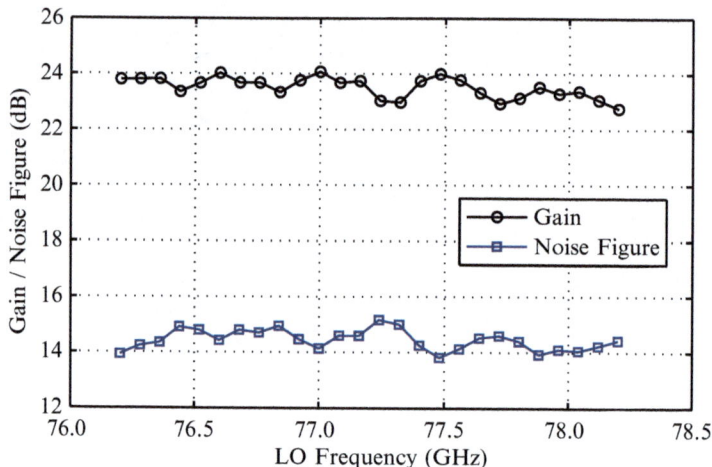

Fig. 5.8 Measurement results of conversion gain and SSB noise figure of the fabricated receiver front-end versus LO frequency with a fixed LO power of 0 dBm and an RF input power of −16 dBm

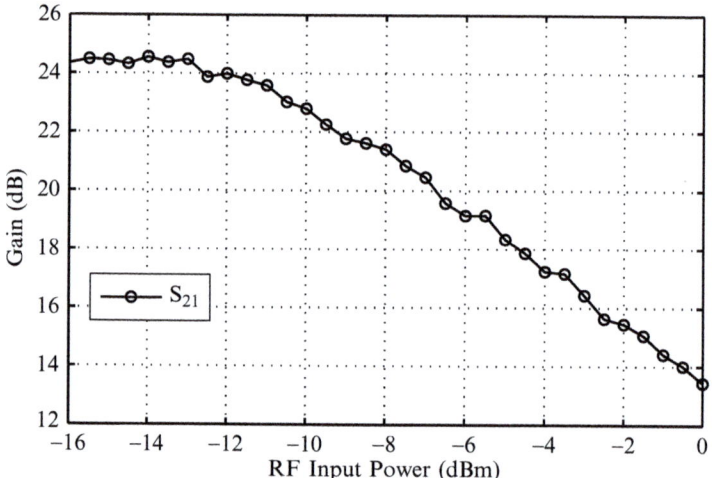

Fig. 5.9 Measurement results of the conversion gain of the fabricated receiver front-end versus RF input power with a fixed LO power of 0 dBm at a frequency of 76.5 GHz

In addition on-wafer S-parameter measurements have been carried out from 10 to 110 GHz using an Agilent network analyzer PNA8361A in combination with waveguide modules for an extended frequency range. Figure 5.10 shows the S-parameter measurements for the LO and RF input ports of the receiver front-end. The input return loss is below −10 dB for both ports in the 77 GHz frequency band.

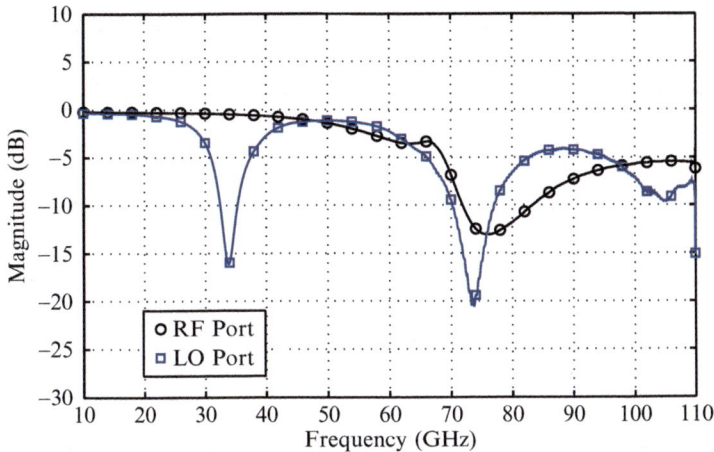

Fig. 5.10 S-parameter measurement results of the LO and RF input return loss for the fabricated receiver front-end

5.4 Conclusion

A novel high-linearity architecture for integrated 77 GHz receiver front-ends in SiGe technology has been presented. The fabricated chip can be operated in differential or single-ended mode and on-wafer measurements show a gain of 24 dB and a noise figure of 14 dB when driven single-ended with an LO power of 0 dBm. An input related 1 dB compression point of -10 dBm is achieved with a total power consumption of 132 mW from a 3.3 V supply. The overall occupied chip area is $728 \times 1028\,\mu m^2$.

Table 5.1 shows a summary of published down-conversion mixers and receiver front-ends in the frequency range of 76–81 GHz in SiGe technology. Equation 5.1 gives the calculation of the figure of merit for the general receiver performance FOM_1. An additional figure of merit FOM_2 that also takes the power consumption and the necessary LO drive into account is given in (5.2).

$$FOM_1 = 174 + Gain - NF + P_{-1dB}(in) \tag{5.1}$$

$$FOM_2 = FOM_1 - P_{DC}(dBm) - P_{LO} \tag{5.2}$$

In comparison to other published receiver front-ends the performance related figure of merit FOM_1 of this work is among the highest published so far. In addition, it simultaneously achieves the best performance to power consumption ratio FOM_2.

Table 5.1 Comparison of SiGe standalone mixers and receivers in the 76–81 GHz band

Ref.	Topology[a]	Gain (dB)	NF[b] (dB)	P_{-1dB}(in) (dBm)	V_{CC} (V)	P_{DC}[c] (mW)	P_{LO} (dBm)	FOM$_1$	FOM$_2$
[1]	Mixer	24	14	−30	5.0	300	2	154	127
[2]	Mixer	11	16.5	−0.3	5.5	412	−3	168	145
[3]	Mixer	13.4	18.4	−12	4.5	176	4	157	131
[4]	Mixer	15.5	16	−3	5.5	187	−2	171	150
[5]	RX	28	11	−16	5.5	1072	1	175	144
[6]	RX + VCO	37	8	−28.5	2.5, 3.5	161		175	
[7]	RX	30	11.5	−26	5.5	440	0	167	140
[8]	RX + VCO	21.7	10.2	−35	5.5	(595)		151	
[9]	RX + VCO	40	6.9	−35	2.5	115		172	
[10]	RX (+ VCO)	40	7–9	−38		122 (195)	−2	168	149
This work	RX	24	14	−10	3.3	132	0	174	153

[a]Brackets denote different published realizations with external and on-chip VCO, respectively
[b]SSB noise figure, DSB noise figures have been increased by 3 dB
[c]Total power consumption without VCO, power consumption including VCO in brackets

References

1. W. Perndl, H. Knapp, M. Wurzer, K. Aufinger, T. F. Meister, J. Böck, W. Simbürger, and A. L. Scholtz, "A low-noise and high-gain double-balanced mixer for 77 GHz automotive radar front-ends in SiGe bipolar technology," in *Proc. IEEE Radio Frequency Integr. Circuits Symp.*, Fort Worth, TX, Jun. 2004, pp. 47–50.
2. B. Dehlink, H.-D. Wohlmuth, H.-P. Forstner, H. Knapp, S. Trotta, K. Aufinger, T. F. Meister, J. Böck, and A. L. Scholtz, "A highly linear SiGe double-balanced mixer for 77 GHz automotive radar applications," in *Proc. IEEE Radio Frequency Integr. Circuits Symp.*, San Francisco, CA, Jun. 2006, pp. 235–238.
3. L. Wang, R. Kraemer, and J. Borngraeber, "An improved highly-linear low-power down-conversion micromixer for 77 GHz automotive radar in SiGe technology," in *IEEE MTT-S Int. Microw. Symp. Dig.*, San Francisco, CA, Jun. 2006, pp. 1834–1837.
4. M. Hartmann, C. Wagner, K. Seemann, J. Platz, H. Jäger, and R. Weigel, "A low-power micromixer with high linearity for automotive radar at 77 GHz in silicon-germanium bipolar technology," in *IEEE Topical Meeting on Silicon Monolithic Integr. Circuits in RF Syst. Dig.*, Long Beach, CA, Jan. 2007, pp. 237–240.
5. B. Dehlink, H.-D. Wohlmuth, K. Aufinger, F. Weiss, and A. L. Scholtz, "An 80 GHz SiGe quadrature receiver frontend," in *IEEE Compound Semicond. Integr. Circuits Symp. Tech. Dig.*, San Antonio, TX, Nov. 2006, pp. 197–200.
6. A. Babakhani, X. Guan, A. Komijani, A. Natarajan, and A. Hajimiri, "A 77 GHz phased-array transceiver with on-chip antennas in silicon: Receiver and antennas," *IEEE J. Solid-State Circuits*, vol. 41, no. 12, pp. 2795–2806, Dec. 2006.
7. M. Hartmann, C. Wagner, K. Seemann, J. Platz, H. Jäger, and R. Weigel, "A low-power low-noise single-chip receiver front-end for automotive radar at 77 GHz in silicon-germanium bipolar technology," in *Proc. IEEE Radio Frequency Integr. Circuits Symp.*, Honolulu, HI, Jun. 2007, pp. 149–152.
8. L. Wang, S. Glisic, J. Borngräber, W. Winkler, and J. C. Scheytt, "A single-ended fully integrated SiGe 77/79 GHz receiver for automotive radar," *IEEE J. Solid-State Circuits*, vol. 43, no. 9, pp. 1897–1908, Sep. 2008.

9. S. T. Nicolson, P. Chevalier, B. Sautreuil, and S. P. Voinigescu, "Single-chip W-band SiGe HBT transceivers and receivers for Doppler radar and millimeter-wave imaging," *IEEE J. Solid-State Circuits*, vol. 43, no. 10, pp. 2206–2217, Oct. 2008.

10. J. Powell, H. Kim, and C. G. Sodini, "SiGe receiver front ends for millimeter-wave passive imaging," *IEEE Trans. Microw. Theory Tech.*, vol. 56, no. 11, pp. 2416–2425, Nov. 2008.

11. D. Kissinger, H. P. Forstner, L. Maurer, R. Lachner, and R. Weigel, "A fully differential low-power high-linearity 77-GHz SiGe receiver frontend for automotive radar systems," in *Proc. IEEE Wireless Microw. Technol. Conf.*, Clearwater, FL, Apr. 2009, pp. 1–4.

12. C. Wagner, M. Treml, M. Hartmann, A. Stelzer, and H. Jäger, "A fully-automated measurement system for 77-GHz mixers," in *Proc. IEEE Instrum. Meas. Techn. Conf.*, Warsaw, Poland, May 2007.

Chapter 6
Differential 77-GHz Current Re-Use Low-Noise Amplifier

6.1 Introduction

The high noise figure of millimeter-wave mixers necessitates the integration of low-noise amplifiers (LNAs) preceding the mixing stage to reduce the overall noise figure of the receiver system. They additionally isolate the LO signal of the mixer from the antenna port to prevent unwanted radiation [1, 2].

The drawback of such an approach is a reduction in linearity due to the additional gain stages introduced through the low-noise amplifier. Cascading of amplifying stages should therefore be kept at a minimum and a small number of transistors with good linearity characteristics and high gain per stage is required.

Published LNAs in the 77 GHz band feature cascaded cascode topologies [3–6] or cascaded common-emitter stages [7–9]. While the cascode configuration suffers from reduced isolation and stability concerns of the common-base stage, the latter necessitates lower supply voltages or dissipates additional unused power.

This chapter presents the design and measurement of 77 GHz narrow-band and 55–77 GHz broadband high-linearity differential LNAs and receivers with high isolation featuring a current re-use stacked common-emitter configuration [10, 11].

6.2 Circuit Design

High-speed transistors possess a maximum collector-emitter breakdown voltage that limits the voltage headroom and subsequently the maximum voltage swing at the output of the transistor. The given supply voltage in a transceiver system which is usually dictated by the integrated oscillator and down-conversion mixer generally exceeds the above breakdown voltage. In a cascaded common-emitter LNA the remaining voltage headroom is dissipated in a stacked resistor. A more efficient way to use the overall headroom is the realization of a cascode topology where

D. Kissinger, *Millimeter-Wave Receiver Concepts for 77 GHz Automotive Radar in Silicon-Germanium Technology*, SpringerBriefs in Electrical and Computer Engineering, DOI 10.1007/978-1-4614-2290-7_6, © Springer Science+Business Media, LLC 2012

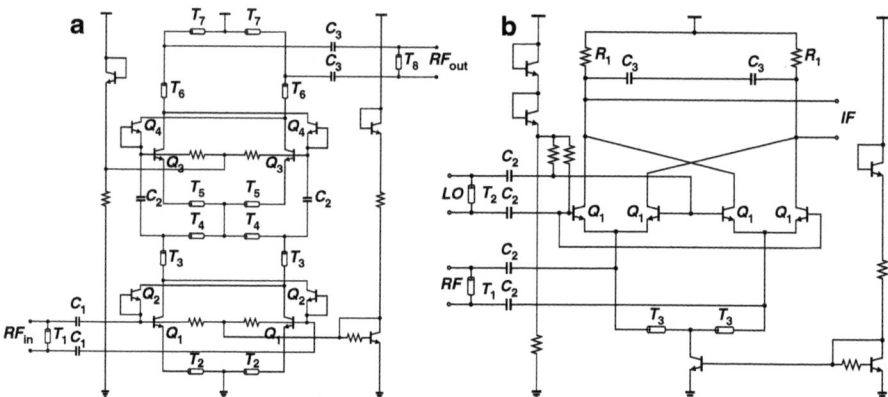

Fig. 6.1 (**a**) Schematic of the proposed differential current re-use vertically stacked common-emitter LNA. (**b**) schematic of the proposed high-linearity mixer

each transistor shares approximately one half of the supply voltage. Nevertheless, the performance of the common-base stage is inferior to the common-emitter configuration in terms of noise, gain, and stability.

Figure 6.1a shows the schematic of the low-noise amplifier including DC bias circuitry. The LNA consist of two differential interstage-matched common-emitter stages Q_1, Q_2 vertically stacked above each other for optimum use of the 3.3 V supply voltage of the overall system. Cross-coupled devices Q_2 and Q_4 are used to improve the reverse isolation of both stages. Due to the limited gain of the first stage, both stages are inductively degenerated through transmission lines T_2, T_5 and optimized for simultaneous noise and power match when driven differentially.

The input matching of the LNA is realized in an L-type with series capacitors C_1 and parallel transmission line T_1. Output matching is facilitated by transmission lines T_6 and T_7 and capacitors C_3 that form a T-type matching network. A virtual ground node between transmission lines T_4 and T_5 acts as the AC ground for the output of the first stage as well as the input of the second. Capacitors C_2 in combination with transmission lines T_3 and T_4 form a T-type interstage matching network between the two common-emitter stages of the LNA.

The proposed receiver consists of a cascade of the above LNA followed by a differentially connected mixer for direct down-conversion of the received signal. Figure 6.1b shows the schematic of the mixer including DC bias. It resembles a double-balanced switching quad formed by transistors Q_1 with L-type matching networks C_1, T_1 for the RF and C_2, T_2 for the LO input port. The transconductance stage of a Gilbert mixer is omitted to enable high voltage swings at the mixer output at low supply voltages. Instead the RF signal path is decoupled from the current source through $\lambda/4$ lines T_3. The IF output is low-pass filtered through the combination of load resistors R_1 with capacitors C_3 and subsequently connected to a differential buffer composed of emitter followers (not shown) that transform the output impedance to $100\,\Omega$ (differential).

6.3   Experimental Results

6.3.1   Narrow-Band Design

Figure 6.2a, b show the layout and the die photograph of the stand-alone LNA. The circuit has been fabricated in a 200 GHz f_t / 250 GHz f_{max} automotive environment certified SiGe bipolar technology with an overall pad limited chip area of $728 \times 728\,\mu m^2$. The LNA draws 24 mA from a 3.3 V supply.

S-parameter measurements have been carried out by two-port on-wafer measurements using an Agilent PNA8361A network analyzer in combination with waveguide modules. Figure 6.4 shows a photograph and a close-up view of the measurement setup for on-wafer S-parameter characterization of the integrated circuits. Figure 6.5 depicts the single-ended S-parameters of the LNA. In single-ended mode, the LNA shows a maximum gain of 12.5 dB and a gain of 12 dB at the operation frequency of 76.5 GHz. Input return loss and reverse isolation are better than −10 and −40 dB in the targeted automotive radar frequency band from 76 to 77 GHz.

Figure 6.6 shows the differential S-parameters of the narrow-band LNA, obtained through measurement of the mixed-mode matrix. Proper termination of the unused ports has been realized by on-chip resistors, which can be removed through laser-fusing. The LNA shows a maximum differential gain of 14 dB and a gain of 12 dB at the operation frequency of 76.5 GHz. Input return loss and reverse isolation are better than −15 and −40 dB in the intended frequency range.

The current consumption of the overall receiver front-end is 54 mA from a supply of 3.3 V (Fig. 6.3). Characterization of the receiver front-end has been carried out by single-ended on-wafer measurements using two Agilent E8257D signal sources in combination with $2\times$ and $3\times$ multipliers respectively.

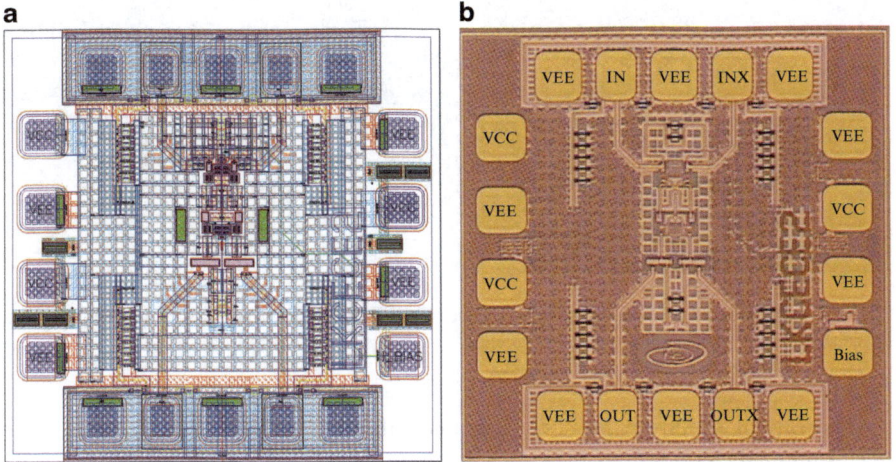

Fig. 6.2 Narrow-band differential current re-use LNA. (**a**) Layout. (**b**) Die photograph

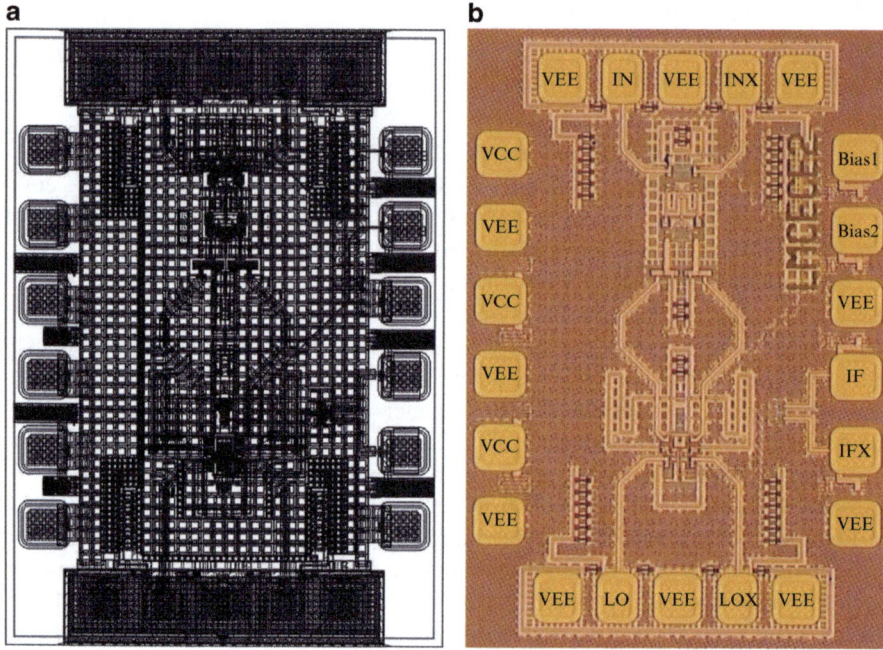

Fig. 6.3 Fabricated receiver front-end. (**a**) Layout. (**b**) Die photograph

Fig. 6.4 S-parameter on-wafer characterization of the fabricated circuits. (**a**) Photograph of the measurement setup. (**b**) Detailed close-up view

The output spectrum has been obtained through a Rohde & Schwarz FSU spectrum analyzer. Gain as well as noise parameters have been determined at an intermediate frequency of 1 MHz.

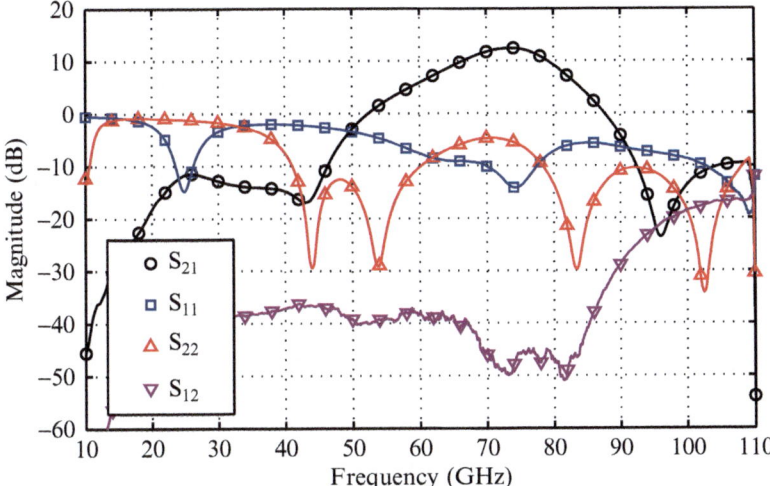

Fig. 6.5 S-parameter measurement results of the fabricated narrow-band current re-use LNA when operated in single-ended mode

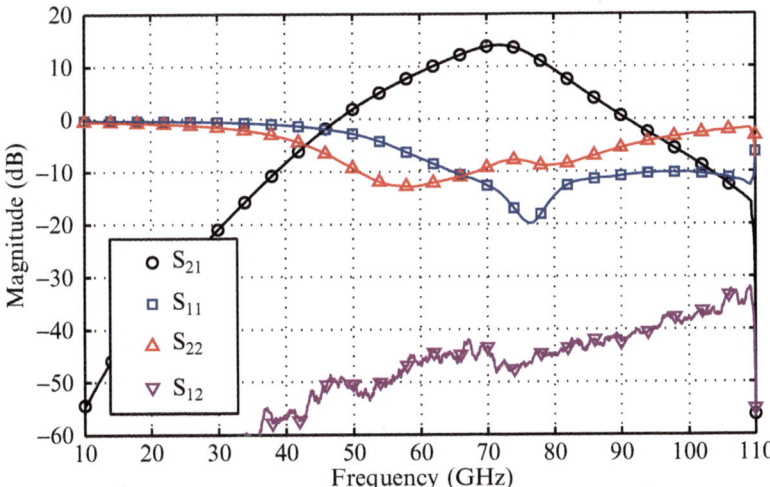

Fig. 6.6 Differential S-parameter measurement results of the fabricated narrow-band current re-use LNA obtained through measurement of the mixed-mode S-parameters. The individual port terminations have been realized through fuseable on-chip resistors

Figure 6.7 shows the variation of noise figure and conversion gain over the local oscillators (LO) power level for a fixed RF input level of -20 dBm. At $+3$ dBm LO power, the receiver achieves a noise figure and conversion gain of 15.8 and 24.5 dB. The measured conversion gain versus the RF input power for a local oscillator power of $+3$ dBm is shown in Fig. 6.8. An input-referred 1 dB compression point of -13 dBm is achieved.

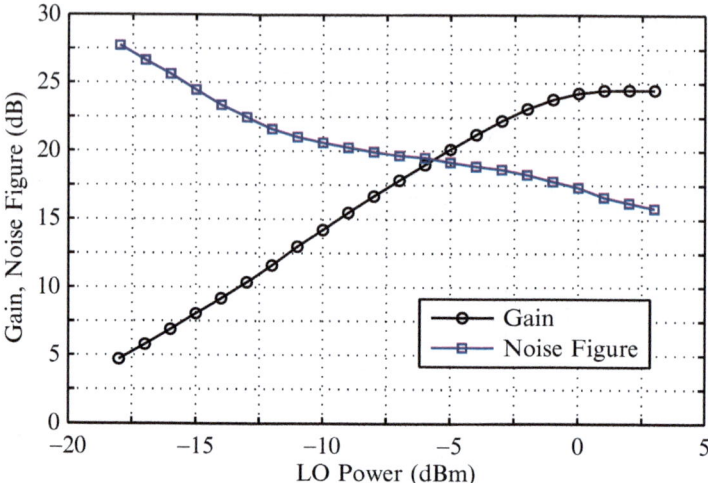

Fig. 6.7 Measurement results of conversion gain and SSB noise figure of the fabricated narrow-band receiver front-end versus applied LO power at a frequency of 76.5 GHz. The RF input power has been fixed to a value of −20 dBm

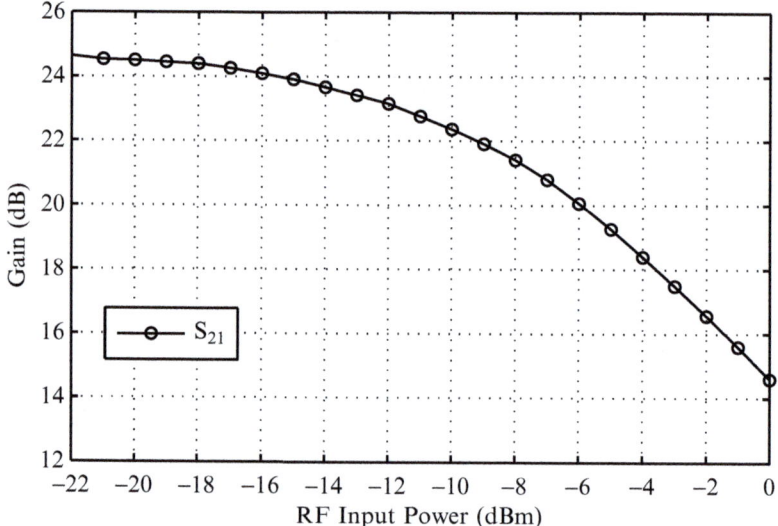

Fig. 6.8 Measurement results of the conversion gain of the fabricated narrow-band receiver front-end versus RF input power with a fixed LO power of +3 dBm at 76.5 GHz

Additional measurements in a similar fashion (not shown) have been carried out for the stand-alone switching quad, yielding a single sideband noise figure of 25 dB. Applying the Friis formula the noise figure of the single-endedly driven LNA can be calculated to 9.5 dB (double sideband). The obtained value agrees well

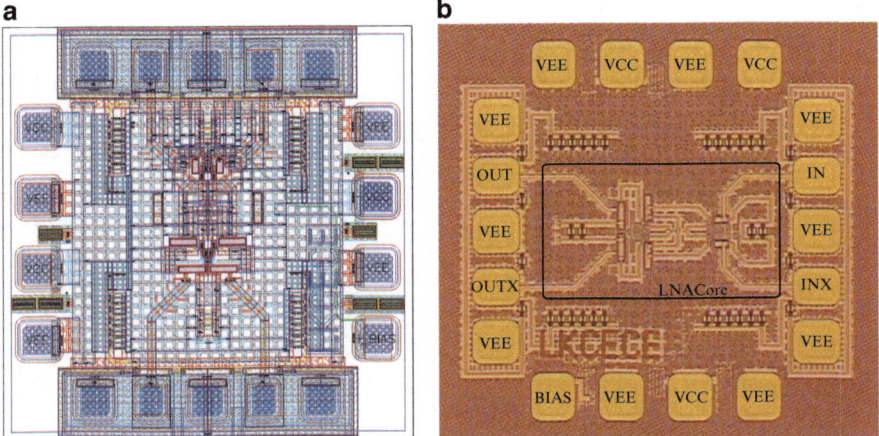

Fig. 6.9 Fabricated broadband differential current re-use LNA in an advanced SiGe technology. (**a**) Layout. (**b**) Die photograph

with simulation results. The LNA has been designed for simultaneous noise and power match in differential mode. Simulations suggest that the noise figure is further improved by 3 dB when the circuit is driven with its optimum noise impedance of $100\,\Omega$ differential. Therefore, a differential noise figure of approximately 6.5 dB can be expected.

6.3.2 Broadband Design

Figure 6.9a, b show the layout and die photograph of the stand-alone LNA. The circuit has been fabricated in a $220\,\text{GHz}\ f_t\,/\,285\,\text{GHz}\ f_{max}$ advanced SiGe bipolar technology. The fabricated LNA draws 12 mA from a 3.3 V supply including bias circuitry and consumes an overall pad limited chip area of $728 \times 728\,\mu\text{m}^2$. The LNA core area as well as the pad configuration have been indicated. The layout and die photograph of the combination of the LNA and the mixing stage used for noise figure determination is depicted in Fig. 6.10. The fabricated receiver front-end draws 42 mA from a 3.3 V supply including bias circuitry with an overall pad limited chip area of $728 \times 1028\,\mu\text{m}^2$.

S-parameter measurements as well as gain and noise figure characterization have been performed with the measurement setup explained in the previous section. Figure 6.11 depicts the single-ended S-parameters of the LNA, while Fig. 6.12 shows the single-ended measurement results for the gain of the amplifier. In single-ended mode, the LNA shows a maximum gain of 13.9 dB at a center frequency of 66 GHz.

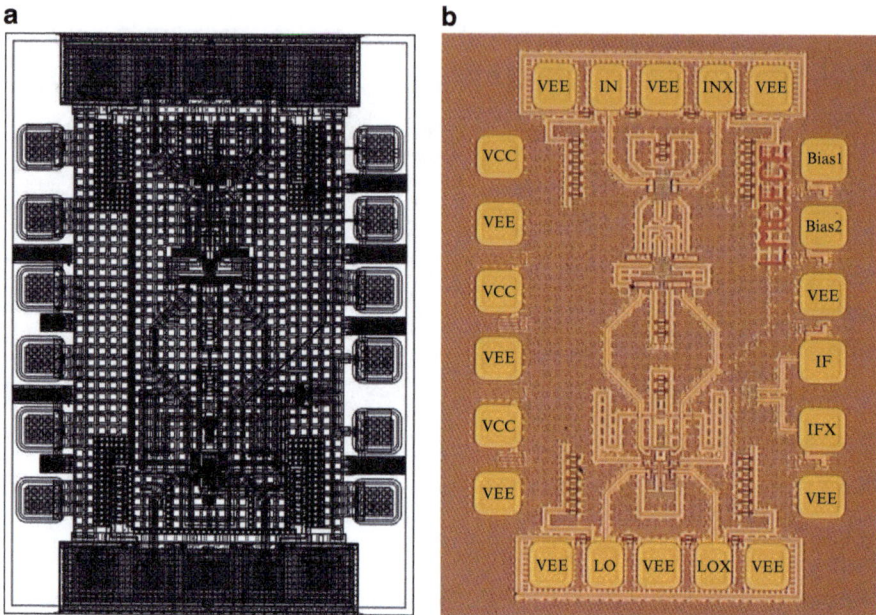

Fig. 6.10 Fabricated receiver front-end consisting of the broadband LNA cascaded with a mixer. (**a**) Layout. (**b**) Die photograph including pad configuration

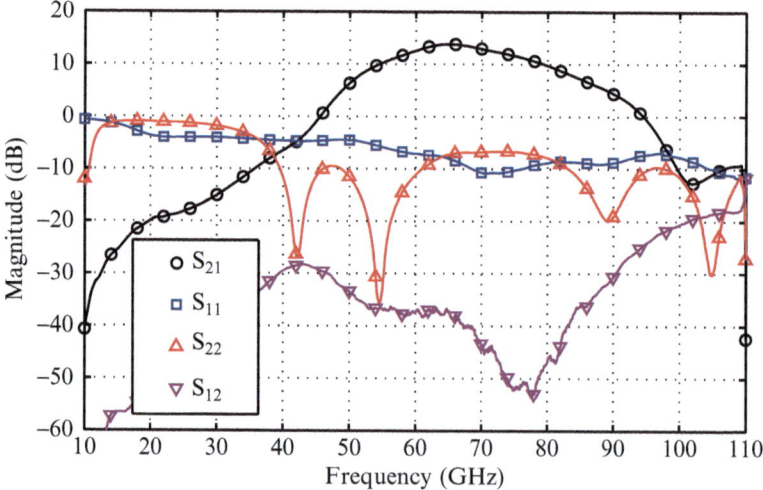

Fig. 6.11 S-parameter measurement results for input and output matching, and isolation of the fabricated broadband current re-use LNA when operated in single-ended mode

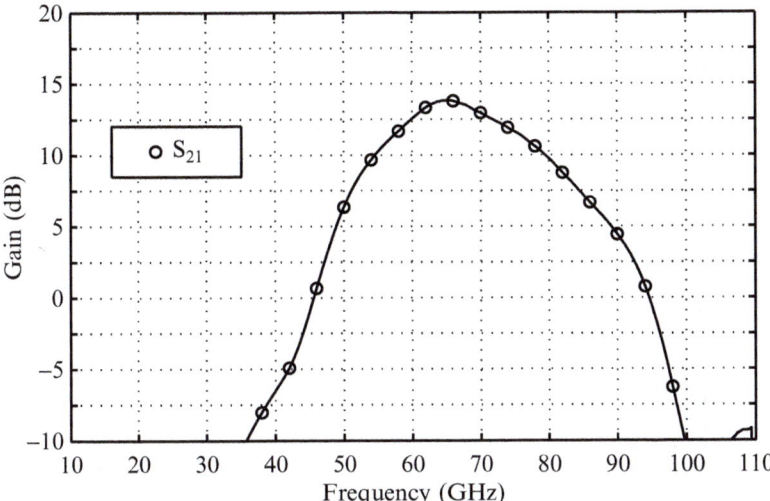

Fig. 6.12 S-parameter measurement results for the gain of the proposed broadband current re-use LNA when operated in single-ended mode

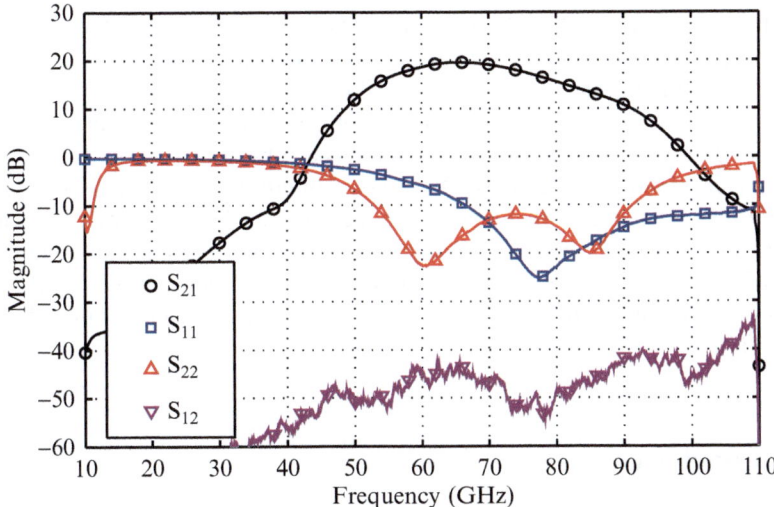

Fig. 6.13 Differential S-parameter measurement results for input/output matching and isolation of the fabricated broadband current re-use LNA obtained through measurement of the mixed-mode S-parameters

Differential S-parameter measurements of the proposed LNA are given in Fig. 6.13. Figure 6.14 shows the differential measurement results for the gain of the amplifier. Both have been computed from the measured mixed-mode parameters. The LNA shows a maximum differential gain of 19.7 dB at the center frequency of

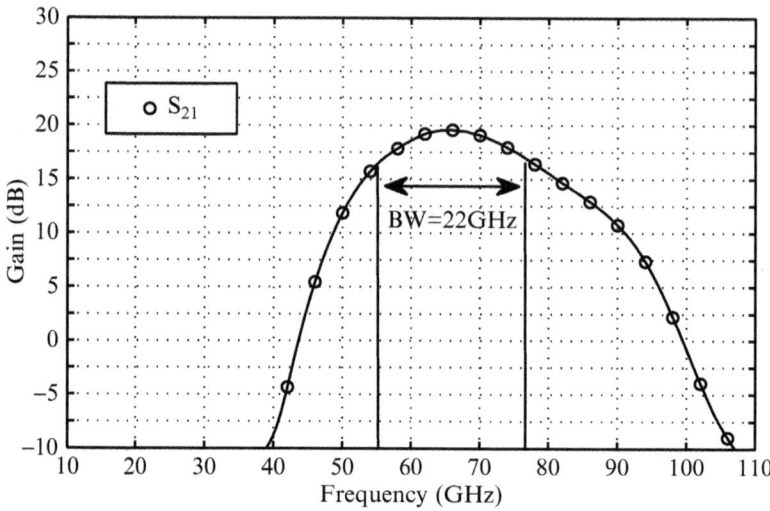

Fig. 6.14 Differential S-parameter results for the gain of the fabricated broadband current re-use LNA obtained through measurement of the mixed-mode S-parameters

66 GHz and a 3-dB bandwidth of 22 GHz. The amplifier is unconditionally stable and the stability factor k remains above four over the entire frequency band.

The overall receiver front-end composed of the broadband LNA and the mixer stage achieves a double-sideband noise figure and conversion gain of 11.4 and 25 dB, respectively. Additional measurements have been carried out for the stand-alone mixer, yielding a double-sideband noise figure of 22 dB. Applying the Friis formula the noise figure of the single-endedly driven LNA can be calculated to 5.8 dB. Additional linearity measurements show an input-referred 1 dB compression point of −14 dBm. Figure 6.15 depicts the measured phase relationship for the transmission of the differential LNA. It shows a constant slope within the 3-dB bandwidth.

6.4   Conclusion

A novel differential current re-use LNA architecture with improved isolation for integrated 77 GHz receivers in SiGe technology has been presented. The fabricated chips can be operated either in differential or single-ended mode.

The narrow-band LNA shows a gain of 12 dB and reverse isolation better than −40 dB in both modes with a differential SSB noise figure around 6.5 dB. An input-related 1 dB compression point of −13 dBm is achieved with a total power consumption of 79 mW from a 3.3 V supply. The occupied chip area is $728 \times 728\,\mu\mathrm{m}^2$.

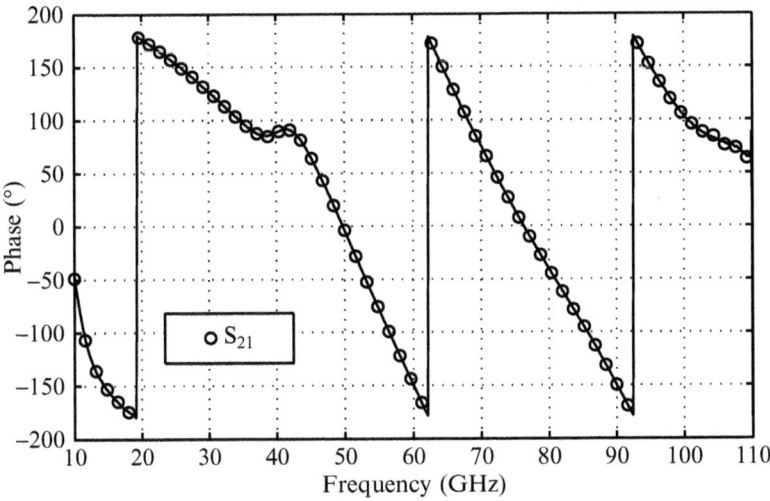

Fig. 6.15 Measurement results of the transmission phase relationship versus frequency of the fabricated differential broadband current re-use LNA

Table 6.1 Comparison of published SiGe low-noise amplifiers in the 76–81 GHz band

Ref.	Topology	No. of Stages[a]	Gain (dB)	NF (dB)	P_{-1dB}(in) (dBm)	V_{CC} (V)	P_{DC} (mW)	Isolation (dB)
[7]	single-ended	2 (CE-CE)	8.9	4.8	−3	5.5	121	−33
[12]	single-ended	3 (CE-CE-CE)	16	6.5 (sim.)		5.5	176	
[3]	single-ended	8 (4x CE-CB)	33	6.2 (sim.)	−40	2.2	81.4	−45
[13]	differential	4 (2x CE-CB)	23.8	5.7		3.5	61.2	
[4]	single-ended	6 (3x CE-CB)	7	8.8 (sim.)	−14	3.0	90	>−40
	differential	6 (3x CE-CB)	13			3.0	90	
[9]	single-ended	3 (CE-CE-CE)	25	5.3 (sim.)	−21	2.5	60	>−50
[5]	single-ended	6 (3x CE-CB)	21.7	10.2 (sim.)	−20	3.5	105	>−45
[6]	differential	4 (2x CE-CB)	18–26	4.9–6.0			55	−45
This	single-ended	2 (CE-CE)	12	9.5	<−13	3.3	79.2	>−40
work	differential	stacked	12	6.5		3.3	79.2	>−40

[a]CE and CB denote common-emitter and common-base stages, respectively

Table 6.1 compares the narrow-band LNA to other published SiGe LNAs in the 76–81 GHz band. The presented LNA shows the highest gain per stage at a comparable noise figure with simultaneous high linearity and low power consumption.

Furthermore, a broadband LNA with 30% fractional bandwidth in SiGe technology has been presented. The LNA shows a differential gain of 19.7 dB at the center frequency of 66 GHz and a noise figure of 5.8 dB at the upper corner frequency of 77 GHz. It dissipates 40 mW from a 3.3 V supply and an output related 1-dB compression point of at least +3 dBm is achieved.

Table 6.2 Comparison of published broadband V-band low-noise amplifiers in silicon

Ref.	Top.[a]	No. of Stages[b]	f_{center} (GHz)	BW (GHz)	BW (%)	Gain (dB)	NF (dB)	P_{1dB} (dBm)	V_{CC} (V)	P_{DC} (mW)
[14]	SE	6 (3CS-CG)	58	14	24	12	8.8	+2	1.5	54
[15]	SE	4 (2CS-CG)	58	15	26	14.6	5.5	−0.5	1.5	24
[16]	SE	3 (3CS)	58	6	10	15	4.4	−4	1.3	7.8
[17]	SE	2 (CS-CG)	54	12	22	11.5	4.1	−5	1.5	16.5
[18]	SE-D	5 (3CS 2CG)	60	7.7	13	19.3	6.1	+2.7	1.2	35
[19]	diff.[c]	2 (2CS)	61	23	37	10	3.8	−4.6	1.2	35
[20]	SE	3 (2CB 1CE)	61	11	18	14.7	4.5	−5.3	1.8	11
[21]	SE	4 (2CE-CB)	60	9	15	14	6 (sim.)	–	3.3	36
[22]	SE	4 (2CE-CB)	62	16	26	15.8	6.8	–	2.5	24
This work	diff.	2 (2CE) stacked	66	22	30	19.7	5.8	+3	3.3	40

[a]*SE* single-ended, *diff* differential, *SE-D* single-ended to differential
[b]*CS* common-source, *CG* common-gate, *CE* common-emitter, *CB* common-base
[c]de-embedded results

Table 6.2 compares the work to other published V-band LNAs in silicon technology. The presented broadband LNA performs favorable against other works. It shows the highest gain per stage, the highest gain-bandwidth product, as well as the highest output-referred 1-dB compression point at comparable power consumption.

References

1. A. Chakraborty, "Design of a 120 GHz broadband LNA in SiGe technology," Master's thesis, Dept. of High Freq. Electron., Tech. Univ. Darmstadt, Darmstadt, Germany, 2009.
2. A. Chakraborty, H. L. Hartnagel, D. Kissinger, B. Laemmle, and R. Weigel, "Design of gain optimized broadband low noise amplifiers at 120 GHz using SiGe technology," in *IEEE German Microw. Conf.*, Berlin, Germany, Mar. 2010, pp. 268–271.
3. R. Reuter and Y. Yin, "A 77 GHz (W-band) SiGe LNA with 6.2 dB noise figure and gain adjustable to 33 dB," in *Proc. Bipolar/BiCMOS Circuits Technol. Meeting*, Maastricht, The Netherlands, Oct. 2006.
4. S. Chartier, B. Schleicher, F. Korndörfer, S. Glisic, G. Fischer, and H. Schuhmacher, "A fully integrated fully differential low-noise amplifier for short range automotive radar using a SiGe:C BiCMOS technology," in *Proc. Eur. Microw. Integr. Circuits Conf.*, Munich, Germany, Oct. 2007, pp. 407–410.
5. L. Wang, S. Glisic, J. Borngräber, W. Winkler, and J. C. Scheytt, "A single-ended fully integrated SiGe 77/79 GHz receiver for automotive radar," *IEEE J. Solid-State Circuits*, vol. 43, no. 9, pp. 1897–1908, Sep. 2008.
6. J. Powell, H. Kim, and C. G. Sodini, "SiGe receiver front ends for millimeter-wave passive imaging," *IEEE Trans. Microw. Theory Tech.*, vol. 56, no. 11, pp. 2416–2425, Nov. 2008.
7. B. Dehlink, H.-D. Wohlmuth, K. Aufinger, T. F. Meister, J. Böck, and A. L. Scholtz, "A low-noise amplifier at 77 GHz in SiGe:C bipolar technology," in *IEEE Compound Semicond. Integr. Circuits Symp. Tech. Dig.*, Palm Springs, CA, Nov. 2005, pp. 287–290.

8. B. Dehlink, H.-D. Wohlmuth, K. Aufinger, F. Weiss, and A. L. Scholtz, "An 80 GHz SiGe quadrature receiver frontend," in *IEEE Compound Semicond. Integr. Circuits Symp. Tech. Dig.*, San Antonio, TX, Nov. 2006, pp. 197–200.

9. S. T. Nicolson, K. A. Tang, K. H. K. Yau, P. Chevalier, B. Sautreuil, and S. P. Voinigescu, "A low-voltage 77-GHz automotive radar chipset," in *IEEE MTT-S Int. Microw. Symp. Dig.*, Honolulu, HI, Jun. 2007, pp. 487–490.

10. D. Kissinger, H. P. Forstner, H. Jäger, L. Maurer, and R. Weigel, "A differential 77-GHz receiver with current re-use low-noise amplifier in SiGe technology," in *IEEE Topical Meeting on Silicon Monolithic Integr. Circuits in RF Syst. Dig.*, New Orleans, LA, Jan. 2010, pp. 128–131.

11. D. Kissinger, K. Aufinger, T. F. Meister, L. Maurer, and R. Weigel, "A high-linearity broadband 55–77 GHz differential low-noise amplifier with 20 dB gain in SiGe technology," in *Proc. Asia-Pacific Microw. Conf.*, Yokohama, Japan, Dec. 2010, pp. 1501–1504.

12. B. Dehlink, H.-D. Wohlmuth, H.-P. Forstner, H. Knapp, S. Trotta, K. Aufinger, T. F. Meister, J. Böck, and A. L. Scholtz, "A highly linear SiGe double-balanced mixer for 77 GHz automotive radar applications," in *Proc. IEEE Radio Frequency Integr. Circuits Symp.*, San Francisco, CA, Jun. 2006, pp. 235–238.

13. A. Babakhani, X. Guan, A. Komijani, A. Natarajan, and A. Hajimiri, "A 77 GHz phased-array transceiver with on-chip antennas in silicon: Receiver and antennas," *IEEE J. Solid-State Circuits*, vol. 41, no. 12, pp. 2795–2806, Dec. 2006.

14. C. H. Doan, S. Emami, A. M. Niknejad, and R. W. Brodersen, "Millimeter-wave CMOS design," *IEEE J. Solid-State Circuits*, vol. 40, no. 1, pp. 144–155, Jan. 2005.

15. T. Yao, M. Q. Gordon, K. K. W. Tang, K. H. K. Yau, M.-T. Yang, P. Schvan, and S. P. Voinigescu, "Algorithmic design of CMOS LNAs and PAs for 60-GHz radio," *IEEE J. Solid-State Circuits*, vol. 42, no. 5, pp. 1044–1057, May 2007.

16. E. Cohen, S. Ravid, and D. Ritter, "An ultra low power LNA with 15dB gain and 4.4dB NF in 90nm CMOS process for 60 GHz phase array radio," in *Proc. IEEE Radio Frequency Integr. Circuits Symp.*, Atlanta, GA, Jun. 2008, pp. 61–64.

17. I. Haroun, J. Wright, C. Plett, A. Fathy, and Y.-C. Hsu, "A V-band 90-nm CMOS low-noise amplifier with modified CPW transmission lines for UWB systems," in *IEEE Topical Meeting on Silicon Monolithic Integr. Circuits in RF Syst. Dig.*, New Orleans, LA, Jan. 2010, pp. 368–371.

18. C. Weyers, P. Mayr, J. W. Kunze, and U. Langmann, "A 22.3dB voltage gain 6.1dB NF 60GHz LNA in 65nm CMOS with differential output," in *IEEE Int. Solid-State Circuits Conf. Dig. Tech. Papers*, San Francisco, CA, Feb. 2008, pp. 192–193.

19. E. Janssen, R. Mahmoudi, E. van der Heijden, P. Sakian, A. de Graauw, R. Pijper, and A. van Roermund, "Fully balanced 60 GHz LNA with 37% bandwidth, 3.8 dB NF, 10 dB gain and constant group delay over 6 GHz bandwidth," in *IEEE Topical Meeting on Silicon Monolithic Integr. Circuits in RF Syst. Dig.*, New Orleans, LA, Jan. 2010, pp. 124–127.

20. B. A. Floyd, S. K. Reynolds, U. R. Pfeiffer, T. Zwick, T. Beukema, and B. Gaucher, "SiGe bipolar transceiver circuits operating at 60 GHz," *IEEE J. Solid-State Circuits*, vol. 40, no. 1, pp. 156–167, Jan. 2005.

21. Y. Sun, F. Herzel, J. Borngräber, and R. Kraemer, "60 GHz receiver building blocks in SiGe BiCMOS," in *IEEE Topical Meeting on Silicon Monolithic Integr. Circuits in RF Syst. Dig.*, Long Beach, CA, Jan. 2007, pp. 219–222.

22. A. Chen, H.-B. Liang, Y. Baeyens, Y.-K. Chen, and Y.-S. Lin, "A broadband millimeter-wave low-noise amplifier in SiGe BiCMOS technology," in *IEEE Topical Meeting on Silicon Monolithic Integr. Circuits in RF Syst. Dig.*, Orlando, FL, Jan. 2008, pp. 86–89.

Chapter 7
Millimeter-Wave Built-In Test Concepts

7.1 Direct Built-In Test Architectures

Test equipment is used to determine and validate the performance parameters and functionality of electronic circuits and overall systems. A general block diagram of a direct test system for receivers is depicted in Fig. 7.1.

The system under consideration is denoted as the *Device Under Test* (DUT). An input signal is applied to the DUT and the input and output response of the circuit is measured and compared to the expected properties of the device. Such tests are used to validate performance parameters in the analog and digital domain.

Testing of millimeter-wave integrated transceivers is a complex, time-consuming, and costly task. External tests necessitate expensive, high-performance measurement equipment with recurring calibration procedures. Furthermore, the continuous increase of operational frequencies leads to the fact that the testing of these RFICs is approaching a significant part of the total manufacturing costs and time.

In order to avoid the above mentioned complex test procedures the integration of an embedded test is highly desirable. The introduction of high-frequency silicon-based technologies with transit frequencies above 200 GHz with simultaneous high levels of integration has created opportunities for such cost-effective embedded test solutions. Low-frequency, and therefore, low-cost, test equipment provides the possibility to evaluate functionality and performance of millimeter-wave front-ends. For single-chip integrated systems this issue is further complicated by the lack of access to RF signal nodes between the building blocks of the circuit. Therefore, future transceiver front-ends require additional on-chip test capabilities [1].

In an integrated transceiver RF electronics are present in the front-end, whereas digital circuitry is used for baseband processing. The front-end can be subdivided into the transmit and receive path with their respective performance parameters like maximum output power, sensitivity, and dynamic range. Analog and RF front-end building-blocks must be tested for their functionality and specification like frequency response, linearity, noise power ratio and absolute power levels [2, 3].

D. Kissinger, *Millimeter-Wave Receiver Concepts for 77 GHz Automotive Radar in Silicon-Germanium Technology*, SpringerBriefs in Electrical and Computer Engineering, DOI 10.1007/978-1-4614-2290-7_7, © Springer Science+Business Media, LLC 2012

Fig. 7.1 Block diagram of a
direct receiver test system;
DUT: *Device Under Test*

Fig. 7.2 Block diagram of an
analog direct mixed-signal
integrated test architecture

Fig. 7.3 Block diagram of a
digitally-controlled direct
mixed-signal integrated test
architecture using D/A and
A/D converters to facilitate
test control in the digital
domain

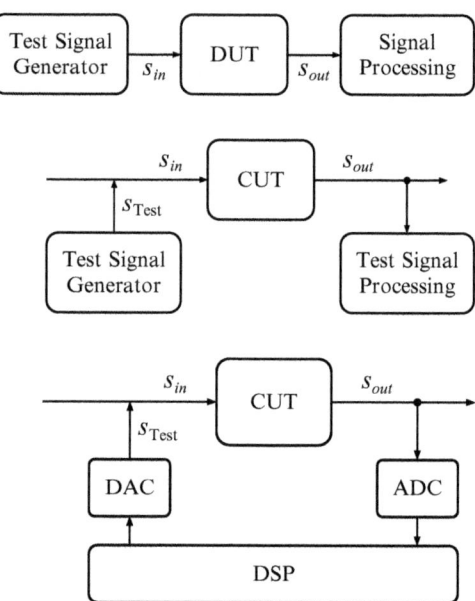

For the baseband-circuitry, digital test techniques already exist. The term *Built-In Self-Test* (BIST) denotes additional integrated circuitry for the purpose of evaluation of the *Circuit Under Test* (CUT).

Testing of analog RF front-end components involves several steps. Figure 7.2 shows a strict analog approach for direct testing of RF building blocks. An input test signal is generated and introduced into the signal path to excite the circuit. The response of the excited analog circuit is measured with a test signal processing unit, e.g. a power detector that operates over a narrow frequency band.

Built-In Self Test (BIST) schemes in mixed-signal circuits make use of integrated analog-to-digital (ADC), digital-to-analog converters (DAC) and DSP blocks for faster and more accurate measurements. Figure 7.3 resembles a possible implementation of such an integrated test. The DSP core generates a sinusoidal input stimulus via the DAC, which is subsequently applied to the input of the analog circuit under test. The output response is converted back into the digital domain by the ADC and analyzed through a fast Fourier transformation (FFT).

When operating frequencies of the front-end circuitry enter the RF or millimeter-wave regime, certain difficulties arise which render the above approaches impractical. The performance of AD and DA-converters in the gigahertz-regime is limited, and sometimes even impossible. As the wavelength of operational frequencies becomes comparable to the chip dimensions, microwave theory has to be applied to the circuit that involves power constraints rather than voltages and currents over time, and is often described by S-parameters and characteristic impedances.

RF devices usually process high-frequency signals that are sinusoidal in nature. That means they have a small bandwidth. Such test signals can be directly generated

Fig. 7.4 Block diagram of a high-frequency direct integrated test architecture. Directional couplers at the input and output are used to introduce and extract the test signal

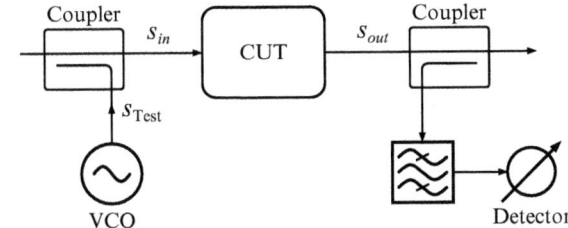

on chip with test oscillators. Another common technique is the use of frequency translating circuits to up-convert a baseband test signal into the RF spectrum of the circuit under test. The input test signals have to be routed into the signal path as depicted in Fig. 7.4. Care has to be taken not to load the regular RF signal path.

7.2   Test Signal Generation

An important aspect for BIST of RF circuitry is the generation of a precision test signal for stimulation of the CUT. Such generators should be small in size, to reduce the overhead in chip area. The test stimuli can be of various types, for example, DC, sinusoid, square wave, or randomly distributed with a known probability distribution. A common aspect of all generated signals is their periodicity. Linear analog circuits are commonly tested with sinusoidal inputs to verify magnitude and phase, as well as harmonic content of the output signal as a function of the input frequency. This principle can be further extended through the use of multi-tone waveforms for analysis of intermodulation characteristics of the CUT. As an alternative to such deterministic spectral-based approaches, probabilistic methods with randomly distributed input signals that have a known spectral distribution can be used to derive the transfer characteristic via the analysis of the spectrum of the response at the output. Such an approach is less direct, but generally easier to implement. ADCs are often characterized in this manner.

Depending on the specific parameters of interest different test signals have to be applied to a receiver. Table 7.1 shows an overview of common test methods used for the performance characterization and validation of integrated receiver front-ends with their respective applied input and obtained output measurands. Except for the Loopback test all approaches require precise calibrated external signal sources.

In a single-tone measurement a signal with a fixed frequency is used to determine the gain of the receiver by measurement of the amplitude or power at the IF output in the linear region. The compression point can be obtained by further increasing the input power until saturation occurs. Through measurement of the noise power density at the output the direct noise measurement method after (7.1) can be used to calculate the noise figure with a 50 Ω termination at the receiver input.

$$NF = N_{out} \, [\text{dBm/Hz}] - G \, [\text{dB}] + 174 \, \text{dBm/Hz} \qquad (7.1)$$

Table 7.1 Overview of different direct test methods for the performance characterization of integrated receiver front-ends with their respective input and output measurands

Input test signal	Measured output signal	Performance parameter
Single-tone	Amplitude/power	Gain, NF, compression
Two-tone	Power spectral density	Gain, NF, compression, IIP2, IIP3
Multi-tone	Power spectral density	Gain, NF, compression, IIP2, IIP3, isolation
Y-factor method	Noise power spectral density	Gain, NF
Loopback test	Internal processing	Functionality test

With the addition of further simultaneously applied test signals additional performance parameters of the receiver can be determined. In a two-tone test a second input signal with a known frequency offset $\Delta f = f_2 - f_1$ is applied to the input of the CUT. This method is used to analyze the second and third-order intermodulation product (IM2, IM3) behavior over varying input powers. A multi-tone test with three test signals additionally allows to obtain the isolation between LO and RF path.

The Y-factor method uses defined noise power levels as test input signals to determine gain and noise figure of the receiver. It relies on the assumption that the noise output power of a device shows a linear behavior as a function of the noise input power. Two different noise temperatures T_H and T_C are applied to the CUT that resemble two input noise powers of:

$$P_{in,H} = kT_H B \qquad (7.2)$$

$$P_{in,C} = kT_C B \,, \qquad (7.3)$$

with the Boltzmann constant k and the bandwidth of the system B. For a calibrated noise source the excess noise ratio (ENR) is defined as:

$$\mathrm{ENR} = \frac{T_H - T_C}{T_0} \qquad \text{with} \quad T_C = T_0 = 290\,\mathrm{K}\,. \qquad (7.4)$$

In both cases, T_H and T_C the ratio Y of the output noise powers is measured and the noise factor F can be determined by (7.5).

$$F = \frac{\mathrm{ENR}}{Y - 1} \qquad \text{with} \quad Y = \frac{P_{out,H}}{P_{out,C}} \qquad (7.5)$$

Furthermore, the gain G of the device can be measured by the ratio of the differences of the noise output powers to the input powers as follows:

$$G = \frac{P_{out,H} - P_{out,C}}{P_{in,H} - P_{in,C}} \approx \frac{P_{out,H}}{P_{in,H}} \qquad \text{for} \quad \mathrm{ENR} \gg 1\,. \qquad (7.6)$$

Analog test signals can be generated by traditional VCOs, e.g. a Colpitts oscillator. The main drawback of this approach is that these circuits are not easily testable and require additional calibration. In the mixed-signal domain, digital signal generators are used because of their testability through standard digital methods.

Built-in test pattern generation requires signal sources with very high frequency resolution and fast frequency switching while a moderate bandwidth is sufficient. Phase-locked loops do not offer both of these attributes and normally require off-chip loop-filters. Direct digital synthesis (DDS) is well known from signal generation for radar systems and is based on the principle of digital phase-accumulation, ROM-based phase-to-amplitude mapping and digital-to-analog conversion. The frequency resolution of a DDS signal is only determined by the width of the phase-accumulator. The spectral content is inferior to a PLL signal and is dominated by spurs in the Nyquist band due to phase truncation, limited ROM resolution, and DAC nonlinearities.

Instead of a DDS circuit, an alternative approach can be used, which makes use of a delta-sigma encoding scheme [4]. Digital signals of N-bit length can be encoded into a 1-bit stream. The power spectral density of the bit stream is composed of defined amplitude tones distributed across the Nyquist interval. This bit stream is subsequently converted into the analog domain by a recovery filter. The required 1-bit pattern can be generated either by a memory-based approach or through a $\Delta\Sigma$-oscillator. The former is realized by 1-bit linear feedback shift registers, while the latter resembles a digital resonator into which a modulator has been inserted. Matching the frequency characteristics of the encoding scheme with the transfer characteristic of the analog filter enables the generation of high-quality test signals.

The transition to higher operational frequencies in the gigahertz-range prohibits the use of direct digital signal generation. In this case, dedicated oscillator circuitry and frequency translating devices have to be used. The main drawback of stand-alone frequency synthesizers is the additional die area consumed by the oscillator as well as the need for frequency stabilizing components (PLLs) to ensure frequency stability of the exciting signal. Heterodyning principles permit the use of low-frequency test signals which can be generated in the aforementioned digital manner. A frequency translating device, e.g. an up-conversion mixer, is used to modulate the local oscillator carrier for the generation of a high-frequency test signal with a precise frequency offset. This methodology is especially useful for characterization of the receiver path of integrated transceiver circuits.

A block diagram of the so-called Loopback approach is shown in Fig. 7.5. It is used for front-end circuits that implement both the transmit as well as the receiver chain onto a single chip [5]. A duplexer is inserted between the transmitter and receiver to connect both parts of the transceiver. In this way, the transmit path acts as a test signal generator for the receiver chain. Additional high-frequency variable attenuators can be added to reduce the signal strength. The Loopback test has the advantage that it requires only a minimum of additional components. On the other hand, it gives little insight regarding the performance and possible failures of the individual RF building-blocks. Further potential issues when integrating on-chip Loopback circuitry involve crosstalk, signal leakage problems, and RF impedance mismatch conditions.

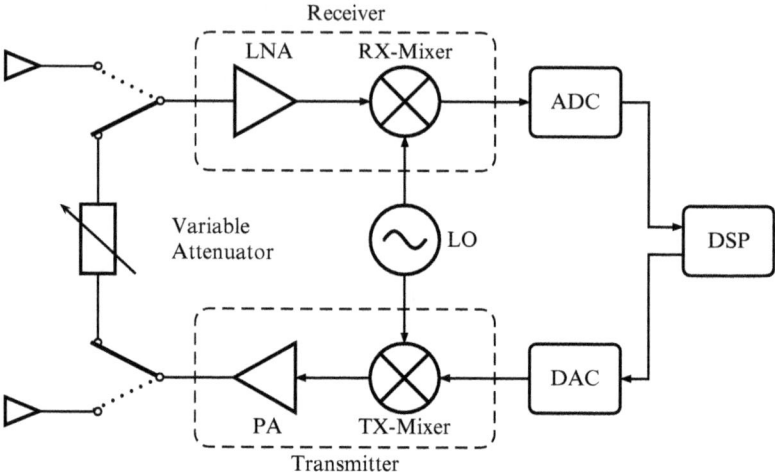

Fig. 7.5 Block diagram of an integrated Loopback-test for a transceiver

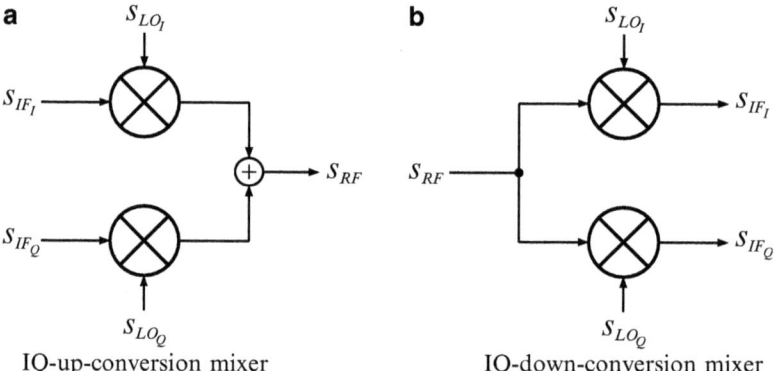

Fig. 7.6 Simplified block diagram of an IQ-mixer. (**a**) IQ-up-conversion mixer for single-sideband generation. (**b**) Corresponding IQ-down-conversion mixer

The use of IQ-modulators for heterodyne generation of the high frequency test signal alleviates the problem of phase dependencies between the LO and the RF double-sideband test signal. It also offers an additional degree of freedom in that it enables the measurement of phase dependencies between the different RF signals.

IQ-mixers suppress the output signal at the mirror frequency through a superposition of orthogonal signals (*In-phase* and *Quadrature*). Therefore, they are also known as *Image-rejection mixers* or *single-sideband mixers*. Figure 7.6 shows a simplified block diagram on an IQ-up-conversion mixer (modulator) and an IQ-down-conversion mixer. They consist of two single mixers that are driven by input signals that are 90° out of phase with respect to each other.

The following calculation are related to an IQ-modulator. For up-conversion into the lower sideband the phase difference of the LO and IF signal has to be of the same sign $(\text{sgn}(\varphi_{LO_q} - \varphi_{LO_i}) = \text{sgn}(\varphi_{IF_q} - \varphi_{IF_i}))$, whereas unequal signs $(\text{sgn}(\varphi_{LO_q} - \varphi_{LO_i}) = -\text{sgn}(\varphi_{IF_q} - \varphi_{IF_i}))$ yield a conversion into the upper sideband.

The modulator input signals:

$$s_{IF_I}(t) = \hat{s}_{IF_I} \cos(\omega_{IF} t) \tag{7.7}$$

$$s_{IF_Q}(t) = \hat{s}_{IF_Q} \sin(\omega_{IF} t) \tag{7.8}$$

$$s_{LO_I}(t) = \hat{s}_{LO_I} \cos(\omega_{LO} t) \tag{7.9}$$

$$s_{LO_Q}(t) = \hat{s}_{LO_Q} (-\sin(\omega_{LO} t)) \tag{7.10}$$

are mixed separately in the in-phase and quadrature path. They are chosen such that mixing into the upper sideband occurs ($\Delta\varphi_{IF} = 90°$; $\Delta\varphi_{LO} = -90°$) yielding the following partial signals in the I- and Q-path:

$$s_{RF_I}(t) = s_{LO_I}(t)s_{IF_I}(t) = \hat{s}_{LO_I} \cos(\omega_{LO} t)\hat{s}_{IF_I} \cos(\omega_{IF} t)$$
$$= \frac{1}{2}\hat{s}_{LO_I}\hat{s}_{IF_I} (\cos(\omega_{LO} t - \omega_{IF} t) + \cos(\omega_{LO} t + \omega_{IF} t)) \tag{7.11}$$

and

$$s_{RF_Q}(t) = s_{LO_Q}(t)s_{IF_Q}(t) = \hat{s}_{LO_Q} (-\sin(\omega_{LO} t)) \hat{s}_{IF_Q} \sin(\omega_{IF} t)$$
$$= \frac{1}{2}\hat{s}_{LO_Q}\hat{s}_{IF_Q} (-\cos(\omega_{LO} t - \omega_{IF} t) + \cos(\omega_{LO} t + \omega_{IF} t)) \ . \tag{7.12}$$

Either of the two signals possesses spectral content in both the upper and lower sideband (double-sideband signal). Nevertheless, while the spectral parts in the upper sideband of both paths are in-phase, the undesired mirror frequency content $\omega_{LO} - \omega_{IF}$ is 180° out of phase between the I and Q-path.

For equal amplitudes of $\hat{s}_{IF_I} = \hat{s}_{IF_Q} = \hat{s}_{IF}$ und $\hat{s}_{LO_I} = \hat{s}_{LO_Q} = \hat{s}_{LO}$ both partial signals can be combined at the output resulting in

$$s_{RF}(t) = s_{RF_I}(t) + s_{RF_Q}(t)$$
$$= \hat{s}_{LO}\hat{s}_{IF} \cos(\omega_{LO} t + \omega_{IF} t) \ . \tag{7.13}$$

The RF output signal does not contain spectral content in the lower sideband due to cancellation of both paths while the signals in the upper sideband are constructively superposed, yielding a single-sideband signal at the output of the modulator.

Imperfections in the phase and amplitude relationships of the IF and LO input signals in implemented modulators cause only partial suppression of the mirror frequency. The image rejection ratio (IRR) denotes the ratio of desired signal

Fig. 7.7 Image rejection ratio (IRR) of an IQ-modulator as a function of the I- and Q-path amplitude ratio k and the phase error φ

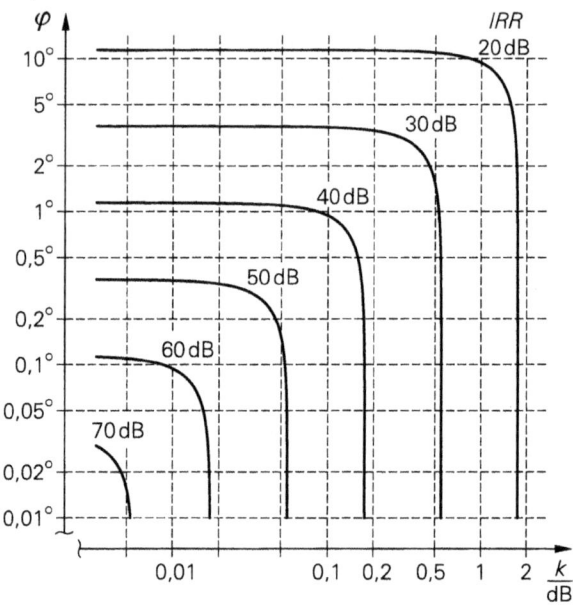

power to the power of the spectral content at the mirror frequency. It represents a performance parameter for IQ-modulators and can be calculated as follows:

$$\text{IRR} = \frac{1+k^2+2k\cos\varphi_f}{1+k^2-2k\cos\varphi_f}.\tag{7.14}$$

In the above equation k denotes the ratio of the output amplitudes of the I and Q-mixer while φ_f describes the phase difference of the two individual paths. Figure 7.7 shows the IRR as a function of amplitude ratio k and the phase error φ_f [6].

7.3 Multiplexing and Coupling

Test signals that have been generated on-chip have to be routed into the signal path. The processed signal then has to be redirected to a detector for evaluation. While this task is easily accomplished in analog and mixed-signal circuits in the voltage domain, it becomes comparably challenging for RF and millimeter-wave integrated circuits. Care has to be taken not to load the signal path, which would directly result in a higher noise figure and lower gain of the receiver path and a reduced maximum power level at the transmitter side, respectively.

Monitoring of RF signal power levels at different nodes for integrated test purposes without affecting the signal path can be accomplished by the insertion of directional couplers. A small portion of the signal is coupled out of the signal

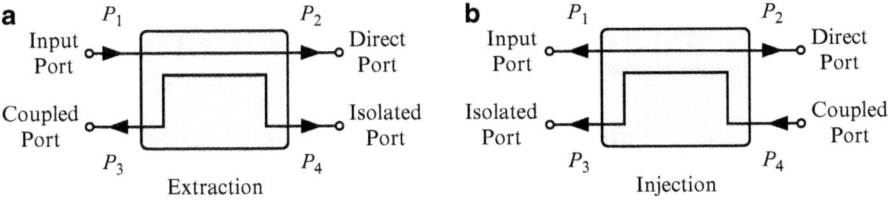

Fig. 7.8 Symbol of a bi-directional coupler with ports and power definitions

path and subsequently measured by a power detector. The DC output voltage of the detector corresponds to the RF power level at the specific node. The power detectors have to work over a wide dynamic range with a very high sensitivity.

Directional couplers in the millimeter-wave regime can be easily integrated as coupled transmission line structures due to the short wavelength used while coupling circuits with lower operational frequencies are usually implemented as lumped passive coupling structures to reduce the overall size of the coupler at the expense of bandwidth [7]. The main drawback of using couplers as routing devices is the high loss of the test signal strength due to the low coupling ratio, which is inversely proportional to the introduced losses in the regular signal path. Figure 7.8 shows the symbol of a bi-directional coupler with ports and power definitions for both extraction and injection of the test signal.

The four ports of the coupler are defined as follows:

Input-Port At the Input-Port the main signal power P_1 is injected.

Direct-Port The Direct-Port forms the main line together with the Input-Port. The major part of the input power P_1 is delivered to this port P_2.

Coupled-Port Depending on the configuration the Coupled-Port can either be used to extract a part of the Input-Port power (a) or to inject test signal power into the main line in the direction of the Direct-Port (b).

Isolated-Port The Isolated-Port is terminated with the characteristic impedance Z_0 for proper operation of the coupler. Alternatively a matched power detector can be connected to the port for additional measurement purposes.

Directional couplers can be characterized by the following performance parameters. In the case below all ports and power definitions correspond to the extracting configuration and proper impedance matching is assumed. Due to the passive nature of the coupler all power ratios (S-parameters) possess a negative sign when given in dB. Nevertheless, the performance parameters are generally given in positive values which is accounted for by an additional negative sign.

Coupling factor The *coupling factor* (*CF*) represents the ratio of extracted power P_3 at the Coupled-Port to injected power P_1 of the main line at the Input-Port.

$$k = -10\log_{10}\frac{P_3}{P_1}\,\mathrm{dB} = -20\log_{10}(S_{31}) = -S_{31_{dB20}} \qquad (7.15)$$

Insertion loss The *insertion loss* (*IL*) specifies the loss in the main line of the directional coupler. It is defined as the ratio of the output power P_2 at the Direct-Port to the input power P_1 of the Input-Port.

$$IL = -10\log_{10}\frac{P_2}{P_1}\,\mathrm{dB} = -20\log_{10}(S_{21})\,\mathrm{dB} = -S_{21_{dB20}} \qquad (7.16)$$

Isolation The *isolation* (*I*) equals the ratio of the output power P_4 at the Isolated-Port related to the injected power at the Input-Port P_1.

$$I = -10\log_{10}\frac{P_4}{P_1}\,\mathrm{dB} = -20\log_{10}(S_{41})\,\mathrm{dB} = -S_{41_{dB20}} \qquad (7.17)$$

Directivity The *directivity* (*D*) represents the ratio of the output power P_4 at the Isolated-Port to output power P_3 at the Coupled-Port related to the input signal P_1 of the main line at the Input-Port.

$$D = -10\log_{10}\frac{P_4}{P_3}\,\mathrm{dB} = -20\log_{10}\frac{S_{31}}{S_{41}}\,\mathrm{dB} = S_{41_{dB20}} - S_{31_{dB20}} \qquad (7.18)$$

One solution to overcome the high losses of the test signal due to couplers is to use multiplexing circuitry. The implementation of RF switches allows for low insertion loss and extraction of the test signal from the signal path. Alternatively, reconfigurable couplers can be used to alter the coupling ratio when the circuit under test is put in test mode. In this case, continuous performance monitoring is not possible for the device. Implementation of RF switches for millimeter-wave circuits is not an easy task. The isolation of switches is limited and introduces additional losses that load the signal path. Another challenge is to maintain proper impedance matching for maximum power transfer as well as avoidance of reflected waves in both the on and off mode of the switching circuitry.

7.4   Test Signal Detection

Analog test signals in the CUT have to be measured and digitized in the integrated circuit for further processing and evaluation. The choice of the measuring element is governed by the operating frequency of the test signal.

While ADCs can be used for the digitization of DC and IF signals high-frequency RF and millimeter-wave signals cannot be sampled in this manner. To evaluate the signal power in the millimeter-wave domain high-performance power detectors are used that convert the measured RF power into a DC or low-frequency value that can subsequently be sampled by an ADC for post-processing in the DSP.

Devices for power measurement are indispensable and standard building blocks for integrated RF and millimeter-wave circuits. Many different circuits are used based on circuit elements in a variety of technologies. The base-emitter junction of

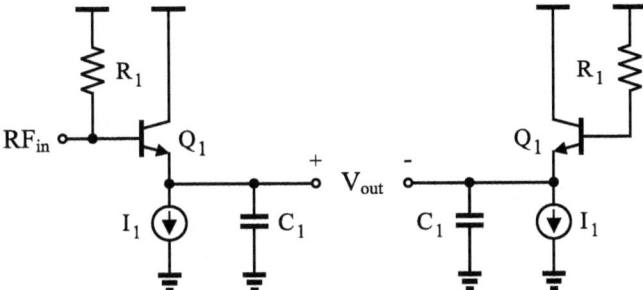

Fig. 7.9 Schematic of a high-frequency peak signal detector circuit with a single-ended input. The power detector features a reference path on the right half for differential output signals

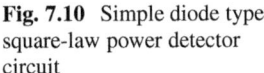

Fig. 7.10 Simple diode type square-law power detector circuit

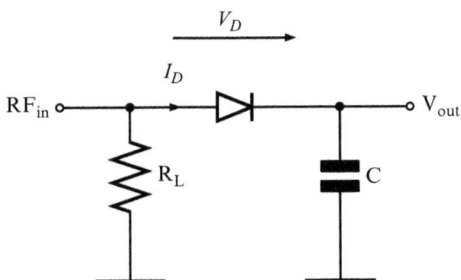

bipolar transistors can be effectively used for microwave power detection, either as peak [8] or as RMS [9] power detectors. A schematic of a peak detector is shown in Fig. 7.9. The single-ended input power detector features a reference path on the right half of the circuit for differential output of the DC power level.

This simple and low-power detector circuit is very attractive for BIST approaches to monitor signal levels or other values. Such detectors can be resistively matched offering very high bandwidths beyond the transit frequency of the transistors. Sensitivity and noise performance can be improved by narrow band conjugate matching, as the power is rectified in the transistor rather than dissipated in the load resistor. The linearity to approximately 0 dBm is sufficient for most measurement cases, but can be improved by cascading detectors with precise attenuators. On the downside, the detector is limited by the noise in the measurement bandwidth. With a 1 Hz measurement bandwidth at the output, signal levels down to −90 dBm can be detected where the output video power is equal to the integrated noise power (noise equivalent power, NEP).

Schottky-barrier diodes for high-frequency performance can be integrated on silicon. Such diodes can be built without process modification in standard CMOS technologies by blocking n^+/p^+ implantation in selected diffusion regions. In modern SiGe BiCMOS technologies these diodes are fabricated using a buried n^+ layer and cobalt salicidation of epi-grown n-type silicon. Figure 7.10 shows the principle of a diode type power detector where the diode is operated as a nonlinear resistor.

The exponential voltage-to-current characteristic of the diode for a positive voltage difference V_D can be described as:

$$I_D(V_D) = I_S \left(e^{\frac{V_D}{nU_T}} - 1 \right) \qquad \text{for} \quad V_D \geq 0. \tag{7.19}$$

In case of high-frequency input signals with a small-signal amplitude $\hat{V}$ the above characteristic can be approximated through a Taylor-series expansion. Excitation of the diode with a signal $v_D(t) = V_0 + \hat{V}\cos(\omega t)$ yields the following expression:

$$\begin{aligned}
i_D(t) &= I_0 + c_1(v_D(t) - V_0) + c_2(v_D(t) - V_0)^2 + \ldots \\
&= I_0 + c_1\hat{V}\cos(\omega t) + c_2\hat{V}^2\cos^2(\omega t) + \ldots \\
&= I_0 + c_1\hat{V}\cos(\omega t) + \frac{1}{2}c_2\hat{V}^2 + \frac{1}{2}c_2\hat{V}^2\cos^2(\omega t) + \ldots .
\end{aligned} \tag{7.20}$$

Equation 7.20 shows that an additional DC part besides I_0 is present in the overall diode current i_D. This component is referred to as the rectified current i_R which is proportional to the square of the voltage amplitude of the input signal and therefore directly related to the input power:

$$i_R = \frac{1}{2}c_2\hat{V}^2 \sim P_{RF} . \tag{7.21}$$

Due to the proportionality of the output current to the square of the input amplitude this type of power detector is also referred to as a *square-law* detector.

References

1. D. Kissinger, B. Laemmle, L. Maurer, and R. Weigel, "Integrated test for silicon front ends," *IEEE Microw. Mag.*, vol. 11, no. 3, pp. 87–94, May 2010.
2. S. S. Akbay, A. Halder, A. Chatterjee, and D. Keezer, "Low-cost test of embedded RF/analog/mixed-signal circuits in SOPs," *IEEE Trans. Adv. Packag.*, vol. 27, no. 2, pp. 352–363, May 2004.
3. G. W. Roberts, "Improving the testability of mixed-signal integrated circuits," in *Proc. IEEE Custom Integr. Circuits Conf.*, Santa Clara, CA, May 1997, pp. 214–221.
4. B. Dufort and G. W. Roberts, "On-chip analog signal generation for mixed-signal built-in self-test," *IEEE J. Solid-State Circuits*, vol. 34, no. 3, pp. 318–330, Mar. 1999.
5. J.-S. Yoon and W. R. Eisenstadt, "Embedded loopback test for RF ICs," *IEEE Trans. Instrum. Meas.*, vol. 54, no. 5, pp. 1715–1720, Oct. 2005.
6. U. Tietze and C. Schenk, *Halbleiter-Schaltungstechnik, (in German)*, 13rd ed. Springer-Verlag, 2010.
7. J. Yoon and W. R. Eisenstadt, "Lumped passive circuits for 5GHz embedded test of RF SoCs," in *IEEE Int. Symp. Circuits Syst.*, Vancouver, Canada, May 2004, pp. 241–244.
8. R. G. Meyer, "Low-power monolithic RF peak detector analysis," *IEEE J. Solid-State Circuits*, vol. 30, no. 1, pp. 65–67, Jan. 1995.
9. T. Zhang, W. R. Eisenstadt, R. M. Fox, and Q. Yin, "Bipolar microwave RMS power detectors," *IEEE J. Solid-State Circuits*, vol. 41, no. 9, pp. 2188–2192, Sep. 2006.

Chapter 8
Direct-Conversion Receiver Built-In Test Architecture

8.1 Introduction

This chapter presents a versatile direct built-in test architecture for integrated radar receiver front-ends [1]. It is used for determination of the gain and noise figure of a direct down-conversion receiver. The test structure is introduced and a detailed analysis of different approaches for the test signal injection and calibration are given.

8.2 Test Structure

The general block diagram of the proposed overall test concept is depicted in Fig. 8.1. For evaluation of the system a test signal has to be generated and routed into the signal path. This test signal can be of various types, e.g. a single-tone sine wave. Measurement of the output signals of both the down-conversion mixer and the power detector is facilitated by the digital signal processing of the transceiver.

By means of a directional coupler the test signal is injected into the signal path of the receiver. A power detector at the isolated port is used to determine the test signal power level. The power spectral density of both the output signals of the power detector and the down-conversion mixer can easily be measured with a low-frequency spectrum analyzer or processed by the ADCs and DSP of the transceiver.

The test signal can either be generated by a dedicated VCO or through a modulation of the LO signal. In the proposed architecture the test signal is generated via a vector modulator. It features up-conversion mixing of the low-frequency test signal in a single sideband. This enables the generation of a single-tone test signal. The use of an SSB modulator is not strictly necessary. The phase dependencies that arise from a double-sideband mixer, e.g. a double-balanced Gilbert cell, can be accounted for by the correct layout. Optimum phase difference between the LO and RF signal can also be achieved through a variable phase shifter in the test signal path.

D. Kissinger, *Millimeter-Wave Receiver Concepts for 77 GHz Automotive Radar in Silicon-Germanium Technology*, SpringerBriefs in Electrical and Computer Engineering, DOI 10.1007/978-1-4614-2290-7_8, © Springer Science+Business Media, LLC 2012

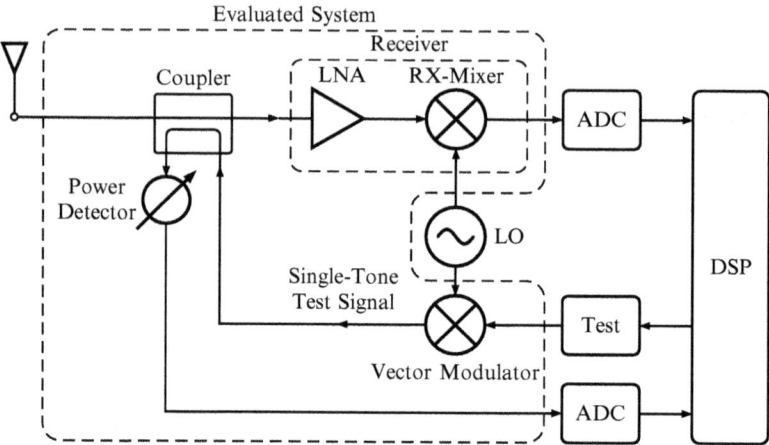

Fig. 8.1 General block diagram of the proposed receiver test concept

The following requirements are given for the test signal generation and injection:

- Single-tone with high SNR and a defined bandwidth,
- Known absolute or relative power levels and defined coupling ratios,
- Low insertion loss in the input signal line of the receiver.

The input power of the test signal P_{in} at the input of the receiver cannot be measured directly. It has to be determined via the output of the power detector and the properties of the directional coupler. The coupled power in the receiver path P_{in} is related to the output power of the vector modulator P_{vecmod} through the coupling factor k:

$$P_{in} = k_{lin}P_{vecmod} \quad \text{with} \quad k_{lin} = 10^{-\frac{k}{10}} . \tag{8.1}$$

The detector senses the power at the isolated port of the coupler P_{det}. This power resembles the vector modulator output power P_{vecmod} reduced by the insertion loss IL of the directional coupler:

$$P_{det} = IL_{lin}P_{vecmod} \quad \text{with} \quad IL_{lin} = 10^{-\frac{IL}{10}} . \tag{8.2}$$

The ratio of the test signal power P_{in} in the receiver path to the power measured at the detector P_{det} can therefore be expressed as follows:

$$P_{in} = \frac{k_{lin}}{IL_{lin}} P_{det} . \tag{8.3}$$

A block diagram of the test signal generator is shown in Fig. 8.2. It is based on a single-sideband vector modulator in which the differential in-phase and quadrature signals are provided by means of a first-order polyphase filter [2].

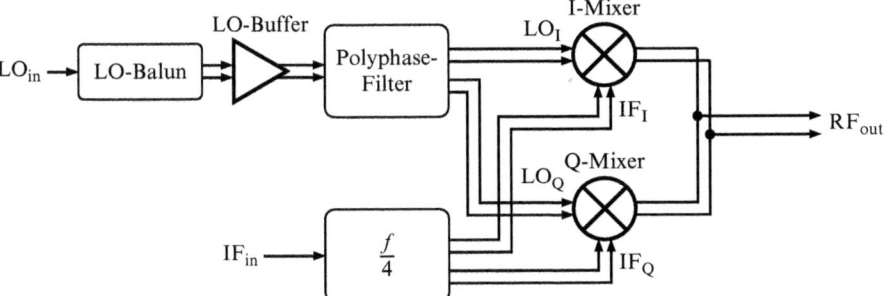

Fig. 8.2 Block diagram of the single-sideband test signal generator. The in-phase and quadrature LO signals are provided by means of a polyphase filter. Generation of low-frequency IF quadrature signals can be realized through a frequency divider by 4

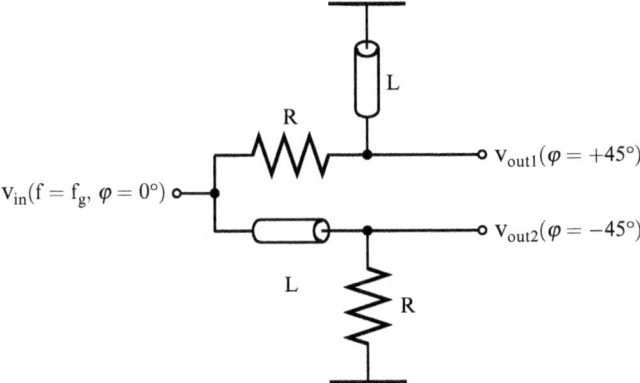

Fig. 8.3 Polyphase filter network for the generation of quadrature signals consisting of a first-order RL high-pass and LR low-pass filter with a phase shift of $\pm 45°$

The polyphase filter is realized as an LR-RL network using integrated transmission lines as inductive elements as depicted in Fig. 8.3. Both the RL high-pass and the LR low-pass filter are designed for a cut-off frequency f_g of 77 GHz yielding a phase shift φ of $+45°$ and $-45°$ at the operational frequency. This results in an overall phase difference of $90°$ between the two output signals v_{out1} and v_{out2}. The insertion loss of each path is 3 dB at the cut-off frequency. An absolute deviation of f_g as well as a mismatch between both the high-pass and low-pass filter results in an amplitude mismatch of the I- and Q-signal which reduces the image rejection.

In the low-frequency domain differential IQ-signals can be generated with the help of static frequency dividers [3]. These dividers are based on D-latches that preserve their previous state or transparently change their output according to the input for a high signal at the clock input (C, CLK). The symbol of a D-latch is given in Fig. 8.4 with the corresponding truth table shown in Table 8.1.

Fig. 8.4 Symbol of a D-latch

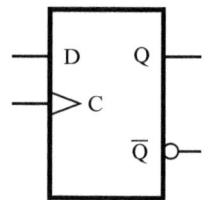

Table 8.1 Truth table of the D-latch

CLK	D	Q
0	0	Q_{-1}
0	1	Q_{-1}
1	0	0
1	1	1

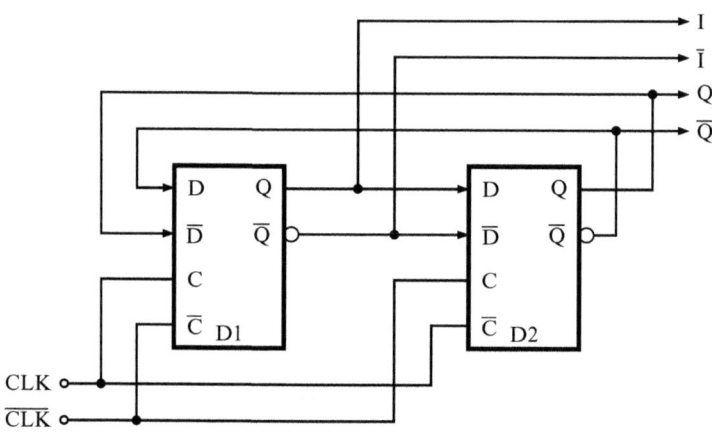

Fig. 8.5 Schematic of a frequency divider with a division ratio of two (divide-by-two) based on differential D-latches with inverted feedback and clock signals

A series connection of two D-latches with inverted feedback and inverted clock signals yields a frequency divider with a division ratio of two (divide-by-two). The resulting schematic of the circuit is shown in Fig. 8.5.

The duty cycle a of the clock signal is defined as the ratio of the high-level time duration τ to the period T of the signal $a = \tau/T$. Signals with a duty cycle of $a = 50\%$ are called symmetric. In the above divide-by-two circuit the frequency of the output signal equals half of the clock input frequency. Unfortunately, the phase difference φ_{IQ} of the I and Q output is dependent on the clock duty cycle a:

$$\varphi_{IQ} = a\pi \tag{8.4}$$

$$= 90° \qquad \text{for} \quad a = 50\%. \tag{8.5}$$

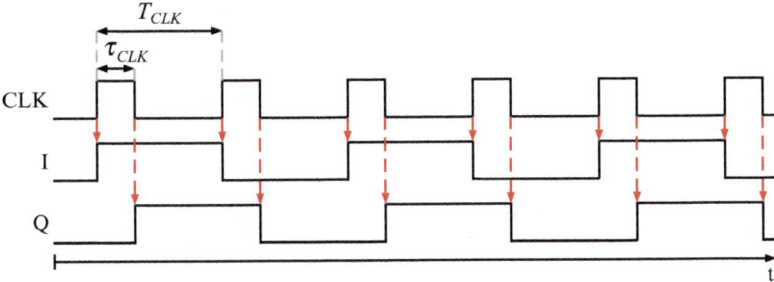

Fig. 8.6 Time response of the I and Q logic levels of the divide-by-two circuit for an unsymmetrical input clock signal (CLK)

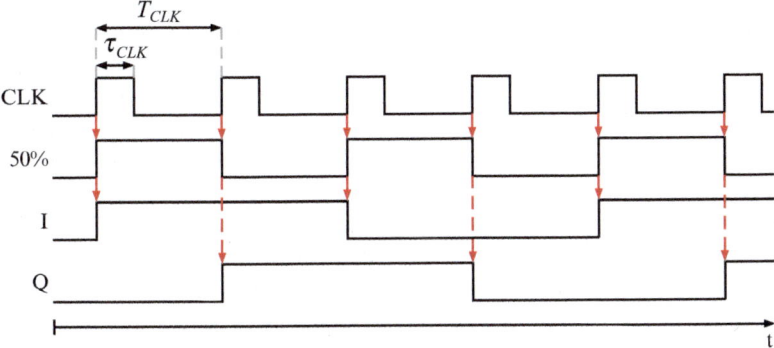

Fig. 8.7 Time response of the I and Q logic levels of the divide-by-four circuit for an unsymmetrical input clock signal (CLK)

As indicated in (8.5) a correct phase difference of 90° can only be achieved for a duty cylce of $a = 50\%$. Figure 8.6 shows the time response of the logic levels of the divide-by-two circuit for an unsymmetrical input clock.

While the phase difference between the I and Q paths is clearly influenced by the symmetry of the clock signal, the duty cycle of the output signals remains constant. It possesses a value of $a = 50\%$ and is only influenced by the jitter of the signal. By cascading two divide-by-two stages a divide-by-four circuit can be realized. The first stage is used to generate a symmetrical signal from the clock at half the input frequency. Subsequently, the second stage performs IQ-signal generation at a quarter of the clock frequency with the required 90° phase relationship. Figure 8.7 shows the respective time response of the logic levels in the divide-by-four circuit for an unsymmetrical input clock.

The rectangular nature of the output signal of the divider results in additional harmonic content at odd multiples of the fundamental output frequency.

A directional coupler is used to inject the test signal into the input path of the receiver. The coupler is realized as a differential $\lambda/4$ microstrip line that operates as a

Fig. 8.8 Block diagram of a differential directional coupler realized through the cross-connection of two differential microstrip transmission lines DL1 and DL2

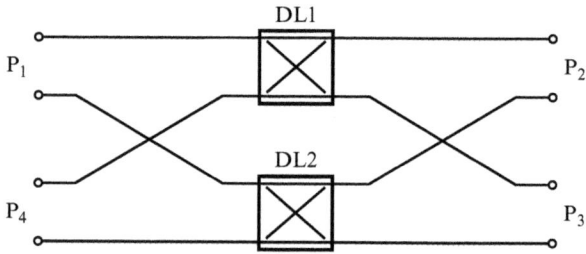

broadside coupled line coupler above a common ground plane. It is implemented in the copper metal stack of the SiGe process featuring SiO_2 as the dielectric material with a relative permittivity of $\varepsilon_r = 3,9$. The distance of the microstrip lines to the ground plane on the bottom metal layer is fixed by the process specification. Through careful optimization of the line width and distance between the coupled structures via EM-simulations a differential microstrip line has been designed that features a single-ended input impedance of $50\,\Omega$ at the individual ports [4]. Figure 8.8 shows the implemented realization of a differential directional coupler through the cross-connection of two differential microstrip transmission lines DL1 and DL2.

For the implementation of the square-law power detector a differential diode-based approach is used. The input of the detector is terminated with a $100\,\Omega$ load and a small fraction of the power is extracted by means of small series capacitors. This eliminates the DC offset of the input signal and reduces the influence of the detector input impedance to the directional coupler. Bipolar transistors are used in a transdiode configuration for the nonlinear power detection. Furthermore, an additional reference path is implemented for cancellation of bias offsets present in the detector circuit. The relationship between the detector output voltage V_{det} and the applied input power P_{in} can be approximated by the following equation:

$$P_{in}\,[\text{dBm}] = f(V_{det}) \approx \frac{p_1 V_{det} + p_2}{V_{det}^2 + q_1 V_{det} + q_2}. \tag{8.6}$$

The above fractional transfer function has to be fitted to a measured dataset of output values of a reference power detector. It can only be applied for a limited range of voltages but proves useful for a quick and accurate estimation of the detector RF input level in the expected power range.

8.3 Analysis

In the following section three different approaches for the determination of gain and noise figure of the receiver are outlined. They mainly differ in the type of test signal provided to the CUT using the architecture proposed in Fig. 8.1. It is important to mention that the overall system is described by a cascade of the directional coupler and the receiver which results in the following system performance parameters:

$$G_{sys} = IL_{lin}G \quad \text{and} \quad F_{sys} = IL_{lin}F \,. \tag{8.7}$$

The first method uses a straight forward approach for the direct determination of both performance parameters [5]. A single-tone low-frequency sinusoidal signal is applied to the test port and subsequently up-converted into the 77 GHz band by the vector modulator. The directional coupler introduces a fraction of the test signal into the receiver input path while the remaining power P_{det} is measured by the power detector. This enables an indirect determination of the test signal input power P_{in} through (8.3) if the parameters of the coupler are known. The IF signal P_{out} at the output of the down-conversion mixer can be measured with an external low-frequency spectrum analyzer or sampled through integrated ADCs followed by a DSP unit. From the ratio of the IF power to the test input power the conversion gain G can be calculated:

$$G = \frac{P_{out}}{P_{in}} = \frac{IL_{lin}P_{out}}{k_{lin}P_{det}} = \frac{IL_{lin}P_{out}}{k_{lin}f(V_{det})} \,. \tag{8.8}$$

The power P_{det} is determined indirectly through the DC voltage V_{det} at the output of the detector. In the linear region of the power detector the inverse transfer function of the detector $P_{det} = f(V_{det})$ can be used. At low test signal levels the SNR at the input of the detector is not sufficient for a precise determination of the input power. On the other hand, a high test signal power level causes saturation effects in the detector. Therefore, the design of the detector has to be optimized for the desired test signal power range.

With the knowledge of the conversion gain G the noise factor F of the integrated receiver can be calculated using the direct noise measurement method (cold source measurement). For that purpose a known noise power density has to be applied to the input of the overall receiver. This can be realized through a 50 Ω termination that is either externally connected to the chip or implemented as a fuseable integrated resistor at the input pads of the device. The power spectral density $kT_0 = -174\,\text{dBm/Hz}$ of the 50 Ω resistor enables the determination of the noise factor F through

$$F = \frac{P_{out}}{kT_0BG} \,. \tag{8.9}$$

From (8.9) it is obvious that the noise factor F is directly related to the output power P_{out}. As in this case the output is present as a noise signal the accuracy of the measurement depends on the resolution bandwidth and the averaging time.

Important for the prediction of performance parameters is the accuracy that is possible with a certain built-in self test. This is strongly dependent on the

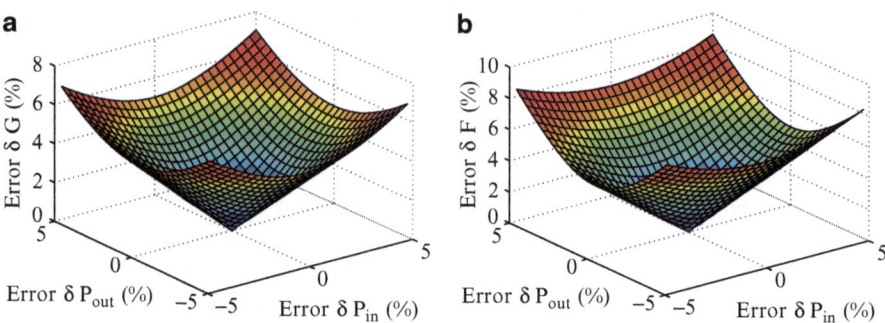

Fig. 8.9 (a) Relative gain error and (b) noise factor error as a function of input and output power measurement accuracy for direct determination of both parameters

measurement method. As every measured value x_m is obtained with a certain error Δx that results in a relative error δx it is imperative:

$$x_m = x + \Delta x$$

$$\delta x = \Delta x / x . \tag{8.10}$$

If a parameter is calculated by the measurement of several values the total error sums up. For stochastically independent measurements it is possible to indicate a most likely error through the application of the Gaussian error propagation:

$$\Delta x = \sqrt{\left(\frac{\partial x}{\partial x_1}\right)^2 \Delta x_1^2 + \left(\frac{\partial x}{\partial x_2}\right)^2 \Delta x_2^2 + \dots} . \tag{8.11}$$

Using (8.11) the overall measurement error that is made in a certain test of performance parameters can be estimated. The accuracy of the proposed measurement method is therefore limited by the error that is made by the power measurement at the input and output power of the test device. It has been shown that the relative error of the receiver gain and noise factor in relation to the relative error of the input and output power measurement can be calculated as follows [5]:

$$\delta G = \triangle G / G = \sqrt{\delta_{P_{in}}^2 + \delta_{P_{out}}^2} \tag{8.12}$$

$$\delta F = \triangle F / F = \sqrt{\delta_{P_{in}}^2 + 2\delta_{P_{out}}^2} . \tag{8.13}$$

Therefore, the accuracy of a predicted noise figure is never as high as the predicted gain, for a direct noise figure measurement. The relative errors dependent on the accuracy of the input and output power are visualized in Fig. 8.9.

The second proposed built-in test approach is based on a general definition of the Y-factor method. For a given noise source that provides two input noise temperatures T_H and T_C to the CUT the resulting output noise powers can be calculated as

$$P_{out,C} = k_B GB(T_C + T_{CUT}) \tag{8.14}$$

$$P_{out,H} = k_B GB(T_H + T_{CUT}). \tag{8.15}$$

With the Y-factor defined as the ratio of the output noise powers,

$$Y = \frac{P_{out,H}}{P_{out,C}} = \frac{T_H + T_{CUT}}{T_C + T_{CUT}}, \tag{8.16}$$

the added noise temperature of the device T_{CUT} can be expressed as

$$T_{CUT} = \frac{T_H - YT_C}{Y-1} = T_C\left(\frac{X-Y}{Y-1}\right) \quad \text{with} \quad X = \frac{T_H}{T_C}. \tag{8.17}$$

which leads to the following solution for the noise factor F:

$$F = 1 + \frac{T_{CUT}}{T_0} = 1 + \frac{T_C}{T_0}\left(\frac{X-Y}{Y-1}\right). \tag{8.18}$$

The above expression enables the general calculation of the noise factor for an input noise source with the power ratio X and a cold noise temperature T_C that can be different from T_0. For the case of $T_C = T_0$ it yields (7.5). A noise figure determination according to (8.18) requires a calibration of T_C which is performed by the regular Y-factor method. Therefore, a $50\,\Omega$ resistor has to be connected to the receiver input. By application of the Y-factor definition of (7.5) to the noise temperatures T_C and T_0 an additional solution for the noise factor F can be given:

$$F = \frac{T_C - T_0}{T_0(Y_C - 1)} \quad \text{with} \quad Y_C = \frac{P_{out,C}}{P_{out,0}}. \tag{8.19}$$

Combination with (8.18) results in the following calibration:

$$T_C = \frac{T_0 Y_C(Y-1)}{(Y-1)-(X-Y)(Y_C-1)} = aT_0 Y_C(Y-1). \tag{8.20}$$

The proposed test procedure provides several advantages over the direct gain and noise figure determination used in the first approach. It solely relies on power ratios for the determination of the noise figure of the receiver. Ratios can generally be measured or generated in integrated circuits with a much higher accuracy than absolute power levels. This proves useful as knowledge of the exact transfer function of the vector modulator and the coupling factor of the directional coupler is not

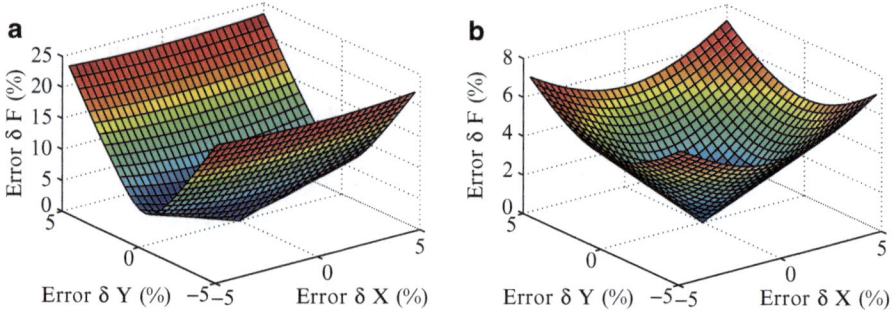

Fig. 8.10 (**a**) Relative noise factor error for $X = 10\,\mathrm{dB}$ and (**b**) $X = 30\,\mathrm{dB}$ versus δY and δX using the Y-factor method applied to a receiver with $NF = 15\,\mathrm{dB}$ and $T_C = T_0$

required. The test signals can be provided by means of a low-frequency noise source with the power ratio X. Furthermore, the approach does not necessitate a power detector at the coupler isolation port which usually suffers from a low accuracy.

For the second test approach a relative noise factor measurement accuracy can be calculated by applying (8.11)–(8.18) which yields

$$\delta F = \frac{T_C}{T_0(Y - 1) + T_C(X - Y)} \sqrt{X^2 \delta X^2 + \left(\frac{Y(1 - X)}{Y - 1}\right)^2 \delta Y^2}. \qquad (8.21)$$

The above expression can be simplified to (8.22) for the special case of $T_C = T_0$. This equation resembles the solution for the common Y-factor method.

$$\delta F = \sqrt{\left(\frac{X}{X - 1}\right)^2 \delta X^2 + \left(\frac{Y}{Y - 1}\right)^2 \delta Y^2} \qquad \text{for} \quad T_C = T_0 \qquad (8.22)$$

A visualization of (8.22) is given in Fig. 8.10. It shows the relative noise factor error δF for two different input noise power ratios X of 10 and 30 dB, respectively. The results have been calculated for a receiver noise figure of $NF = 15\,\mathrm{dB}$ with $T_C = T_0$. As can be seen a high input noise power ratio X is required to prevent a strong dependency on the accuracy of the output power ratio Y.

In the previous treatment a perfect calibration of T_C has been assumed. From (8.20) the relative cold noise temperature error δT_C can be calculated as

$$\delta T_C = a \sqrt{(X(Y - 1))^2 \delta X^2 + \left(\frac{Y(Y_C - 1)(X - 1)}{(Y - 1)}\right)^2 \delta Y^2 + (X - 1)^2 \delta Y_C^2}. \quad (8.23)$$

The relative cold noise temperature error δT_C for a calibration noise power ratio T_C/T_0 of 40 and 50 dB as a function of the input and output power ratio accuracy for $NF = 15\,\mathrm{dB}$ and $X = 30\,\mathrm{dB}$ is depicted in Fig. 8.11. While the dependency on the

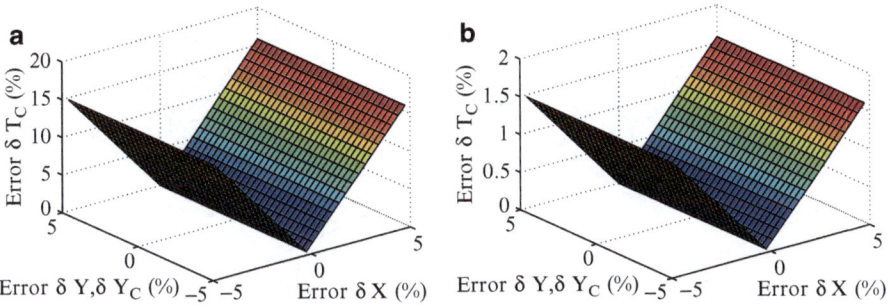

Fig. 8.11 (a) Relative cold noise temperature error for a ratio $T_C/T_0 = 40$ dB and (b) $T_C/T_0 = 50$ dB versus δY and δX for a receiver with $NF = 15$ dB and $X = 30$ dB

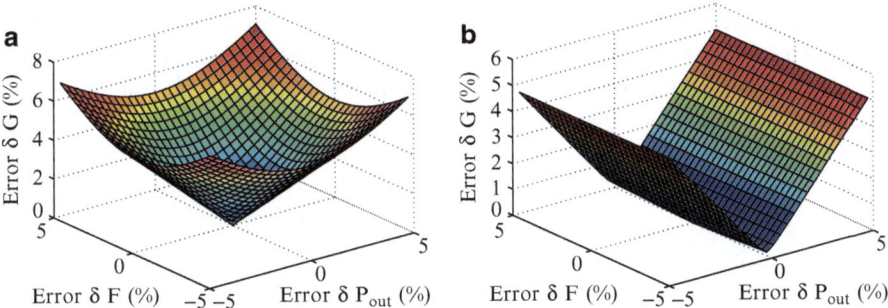

Fig. 8.12 (a) Relative gain error of the receiver for $T_C = T_0$ and $NF = 0$ dB and (b) $T_C/T_0 = 30$ dB with $NF = 15$ dB as a function of δF and δP_{out}

output power ratio accuracy δY is negligible, it can be deduced that a sufficiently high ratio T_C/T_0 is required to suppress the error of the input power ratio X.

For the determination of the receiver gain an absolute measurement of the output power is required. An expression for the calculation of the gain G can be given through combination of (8.14) and (8.18):

$$G = \frac{P_{out,C}}{k_B B(T_C + T_0(F-1))},$$ (8.24)

$$\delta G = \sqrt{\delta P_{out}^2 + \left(\frac{T_0 F}{T_C + T_0(F-1)}\right)^2 \delta F^2}.$$ (8.25)

As shown in Fig. 8.12 for high noise factors the relative gain error δG is dominated by the absolute output power measurement error δP_{out}. A dependency on the error δF can only be observed for low noise factors F in combination with a low calibration noise power ratio T_C/T_0.

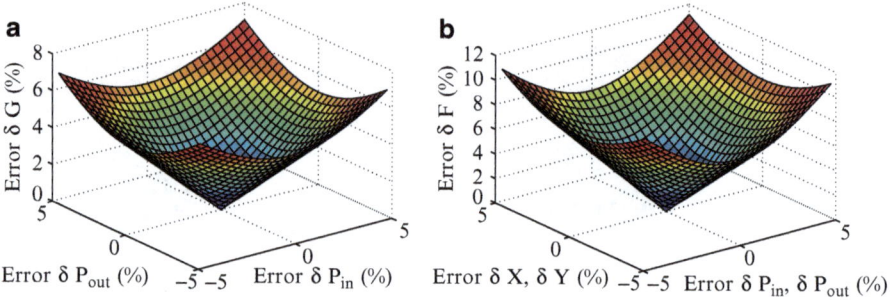

Fig. 8.13 (**a**) Relative gain error and (**b**) noise factor error versus δX, δY and δP_{in}, δP_{out} for a receiver with $NF = 15\,\text{dB}$ and $X = 30\,\text{dB}$ using the third test concept

The third test concept does not require a $50\,\Omega$ termination at the receiver input. The gain of the device is measured by a direct gain measurement with a sinusoidal test signal using the power detector at the directional coupler isolation port. Based on this calibration, the noise factor can be determined through the general Y-factor method [6]. Solving (8.14) for T_C and substitution into (8.18) yields

$$F = 1 + \frac{P_{out,C}}{k_B G B T_0}\left(\frac{X-Y}{X-1}\right), \tag{8.26}$$

$$\delta F = \sqrt{\delta P_{in}^2 + 2\delta P_{out}^2 + \left(\frac{X}{X-1}\right)^2 \delta X^2 + \left(\frac{Y}{Y-1}\right)^2 \delta Y^2}. \tag{8.27}$$

The relative gain and noise factor error dependent on the accuracy of the input and output power δP_{in}, δP_{out} as well as the power ratios δX, δY is shown in Fig. 8.13.

8.4 Conclusion

A versatile built-in test architecture for gain and noise figure determination of direct-conversion radar receivers has been presented. In the following, three different methods for the test signal generation and parameter measurements have been proposed. The direct gain and noise figure method as well as an extended Y-factor method have been described. Furthermore, an error propagation analysis has been performed.

Both of the above concepts necessitate a terminating resistor at the receiver input for calibration and are suitable for circuit verification prior to assembly. Therefore, a third test concept based on a combination of the two approaches has been developed that does not require a noise standard at the input. It can thus additionally be applied as an on-line diagnosis tool during circuit operation in a radar front-end.

References

1. E. Kolmhofer, D. Kissinger, and F. Starzer, "Method and circuit for operating a mixer," U.S. Patent 12/722,511, Mar. 11, 2010.
2. C. Wagner, M. Hartmann, A. Stelzer, and H. Jäger, "A fully differential 77-GHz active IQ modulator in a silicon-germanium technology," *IEEE Microw. Wireless Compon. Lett.*, vol. 18, no. 5, pp. 362–364, May 2008.
3. B. Razavi, *RF Microelectronics*. Prentice Hall, 1998.
4. A. Selz, "Entwurf einer Empfänger-Architektur mit integriertem Testkonzept in SiGe Technologie," diploma thesis, Inst. for Electron. Eng., Univ. of Erlangen-Nuremberg, Erlangen, Germany, 2010.
5. D. Kissinger, R. Agethen, and R. Weigel, "Analysis of a built-in test architecture for direct-conversion SiGe millimeter-wave receiver frontends," in *Proc. IEEE Instrum. Meas. Techn. Conf.*, Austin, TX, May 2010, pp. 944–948.
6. R. Agethen, "Untersuchung integrierter Testkonzepte von 76-81 GHz Radar-Frontends für Automobilanwendungen," diploma thesis, Inst. for Electron. Eng., Univ. of Erlangen-Nuremberg, Erlangen, Germany, 2009.

Chapter 9
Recursive 77-GHz Mixer Test Architecture

9.1 Introduction

Two main drawbacks of integrated test solutions for millimeter-wave receivers exist. The necessity for additional high-frequency test components results in an increased chip area as well as a lower yield which both lead to higher production costs. Furthermore, these supplemental building blocks consume additional power resulting in increased problems regarding heat dissipation of the overall transceiver.

This chapter presents a novel down-conversion mixer architecture which is capable of simultaneous up-conversion of a low-frequency signal in order to generate a high-frequency test signal with a defined frequency offset from the LO carrier. The up-converted double-sideband test signal is coupled back into the mixer input path to enable a built-in functionality test of the receiver path of the mixer. The proposed scheme does not dissipate any additional power for the high-frequency test signal generation and keeps additional consumed chip area at a minimum [1].

9.2 Circuit Design

The proposed recursive architecture for built-in test of single direct down-conversion mixers is depicted in Fig. 9.1a. Both single-ended LO and RF input signals are transformed into differential signals by on-chip LC baluns. The mixer core is driven by an external LO signal and is capable of simultaneous up-conversion of the RF_{in} signal as well as down-conversion of the IF_{in} signal (test signal). The up-converted double-sideband signal is fed back differentially into the RF input path where it is down-converted again so that the input test signal reappears at the IF output. Future realizations will integrate the local oscillator along with the mixer so that no external high-frequency signals will be necessary for functionality test of the circuit. The differential low-frequency test signal is generated externally with a signal source in combination with a low-frequency balun.

D. Kissinger, *Millimeter-Wave Receiver Concepts for 77 GHz Automotive Radar in Silicon-Germanium Technology*, SpringerBriefs in Electrical and Computer Engineering, DOI 10.1007/978-1-4614-2290-7_9, © Springer Science+Business Media, LLC 2012

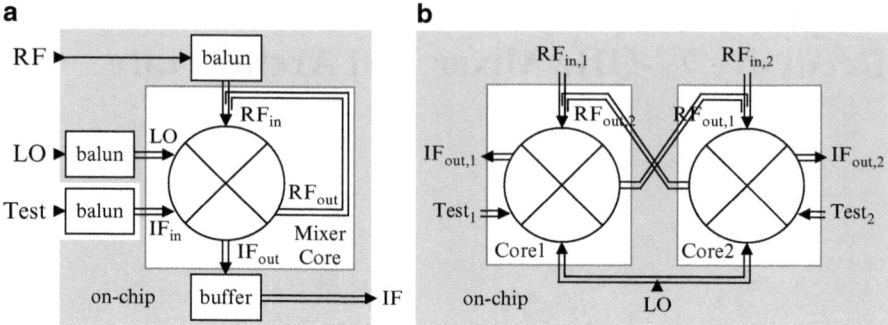

Fig. 9.1 Proposed scheme for a single mixer with built-in functionality test capability. (**a**) Recursive configuration. (**b**) Alternative dual-channel integrated test approach

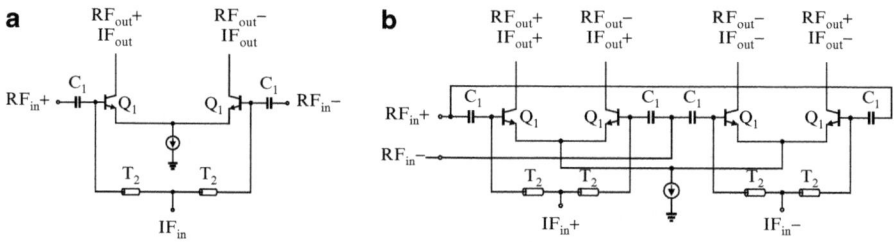

Fig. 9.2 Schematic of a differential amplifier input stage with simultaneous injection of an additional test signal. (**a**) Single-ended. (**b**) Differential

An alternative integrated test approach for a dual-channel receiver architecture is shown in Fig. 9.1b. Both receiver paths feature the above multi-purpose mixer core for simultaneous up- and down-conversion. The mixer core 1 generates a high-frequency test signal $RF_{out,1}$ with a defined frequency offset of $Test_1$ from the LO carrier. This signal is coupled into the receiver path of the second mixer core to enable built-in test of the mixer core 2. Test signal generation for the first mixer core is achieved in a similar fashion through $Test_2$. Both, the proposed recursive and dual-channel architecture, can easily be scaled for multi-channel front-ends.

The principle of simultaneous injection of a test signal into the RF input path of a differential amplifier is shown in Fig. 9.2. According to Fig. 9.2a, a single-ended low frequency test signal IF_{in} can be superimposed on the differential RF signal by use of the virtual ground node between the parallel stubs T_2 as the injection point. Capacitors C_1 and transmission lines T_2 form a matching network for the RF input and additionally act as a frequency separating filter. Figure 9.2b shows a modified version for differential injection of the IF_{in} test signal. The transistors Q_1 have to be separated into two parallel transistors of half the original emitter length. The same method has to be applied for the capacitors C_1 and transmission lines T_2. In this way

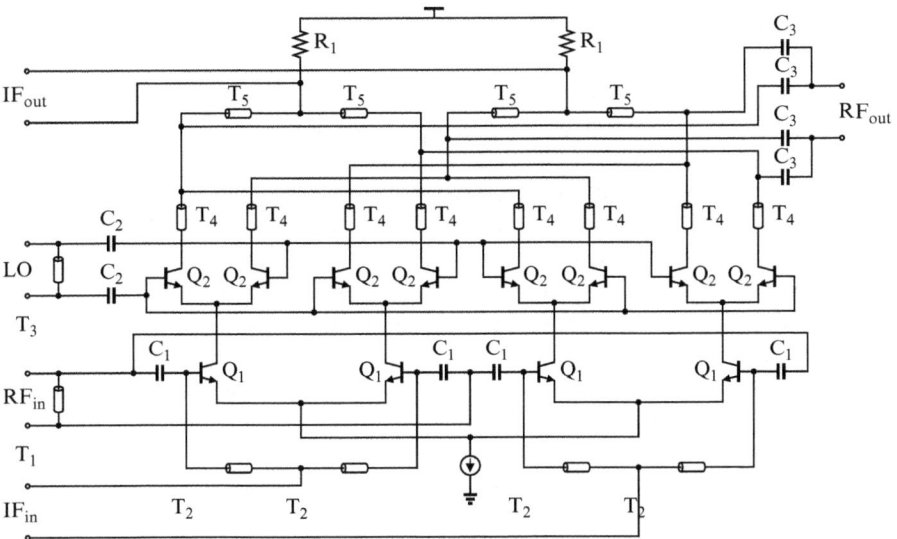

Fig. 9.3 Simplified schematic of the proposed differential mixer core. Bias circuitry is not shown for simplicity

the two differential input signals RF_{in} and IF_{in} appear as common-mode signals to each other. The aforementioned method can also be applied to two high-frequency signals when accounting for the virtual grounds on two different symmetry lines [2].

Figure 9.3 shows a simplified schematic of the proposed overall mixer architecture. It resembles two duplicated symmetric double-balanced Gilbert cell mixers. Both differential inputs RF_{in} and IF_{in} are split up and fed to the transconductance stage transistors Q_1 in a way that they appear as common-mode signals to each other and are superimposed at the four outputs of the transconductance stage. Each of the two differential inputs is using all four transistors Q_1 available. The capacitors C_1 act as an open for the low-frequency input whereas transmission lines T_1 and T_2 represent a low-frequency shorted connection. Transmission lines T_1, T_2, and capacitors C_1 additionally serve as a π-type input matching network for the RF_{in} port.

The double switching quad transistors Q_2 of the mixer core perform simultaneous up- and down-conversion of the amplified input signals. The superimposed output signals are separated again by transmission lines T_4, T_5, and capacitors C_3 so that the up-converted IF_{in} signal (test) appears at RF_{out}, while the regular down-converted RF_{in} signal appears at IF_{out}. The RF_{out} signal is fed back to the RF_{in} port via a differential coupled transmission line coupler with a coupling ratio of approximately -25 dB. The output signal IF_{out} is connected to a differential common-emitter output buffer (not shown) that transforms the output impedance to $100\,\Omega$ differential.

9.3 Experimental Results

Figure 9.4a shows the die photograph of the fabricated mixer with built-in test functionality. A close-up view of the realized mixer core is depicted in Fig. 9.4b. The overall chip area is $1028 \times 1128\,\mu m^2$ and the circuit draws $22\,mA$ from a single $3.3\,V$ supply, including all integrated bias circuitry. The chip has been fabricated in a $200\,GHz$ f_t/$250\,GHz$ f_{max} automotive environment certified SiGe bipolar technology. Different stages as well as the pad configuration of the integrated circuit are indicated in the figure. The number of virtual ground nodes requires careful layout of the mixer core with respect to the two different symmetry lines. The arrangement of the transistors has been optimized to minimize crossover of transmission lines to reduce coupling effects. Integrated baluns have been realized as $\pm 90°$ LC phase shift structures.

Characterization of the integrated recursive mixer has been carried out by single-ended on-wafer measurements using two Agilent E8257D signal sources in combination with $2\times$ and $3\times$ multipliers for the LO and RF path, respectively. The output spectrum has been obtained through a Rohde & Schwarz FSU spectrum analyzer following an external buffer board. Figure 9.5 shows a photograph and a close-up view of the measurement setup for on-wafer characterization of the integrated circuit.

The obtained output spectrum at the IF_{out} port with the RF_{in} port terminated by on-chip $50\,\Omega$ resistors is shown in Fig. 9.6. Test signals with a peak voltage amplitude of $1\,mV$, $3\,mV$, and $10\,mV$ at a frequency of $1\,MHz$ have been applied to the test port (IF_{in}). With the LO input power set to $0\,dBm$ at a frequency of $76.5\,GHz$, the IF output power increases to $-20\,dBm$ for a test signal peak voltage of $10\,mV$. Additionally, higher order harmonic content of the up- and down-converted recursive test signal can be monitored and evaluated at $2\,MHz$ offset frequency.

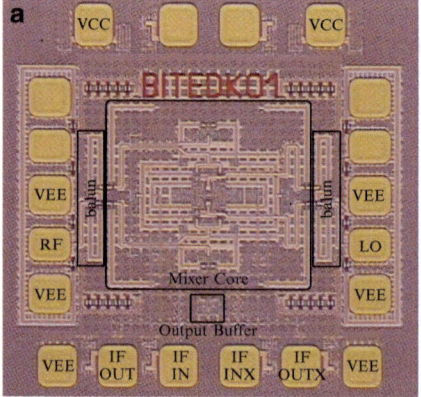 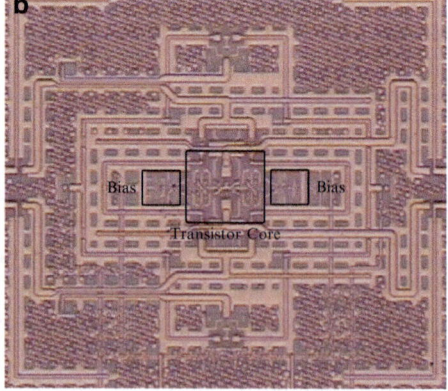

Fig. 9.4 (**a**) Die photograph of the fabricated recursive built-in test down-conversion mixer. (**b**) Close-up view of the mixer core with indicated building blocks

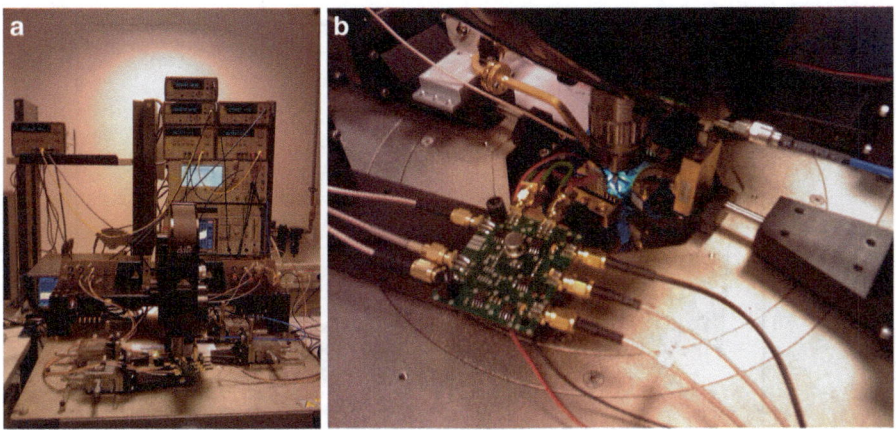

Fig. 9.5 Conversion gain and noise figure on-wafer characterization of the fabricated circuit. (**a**) Photograph of the measurement setup. (**b**) Detailed close-up view

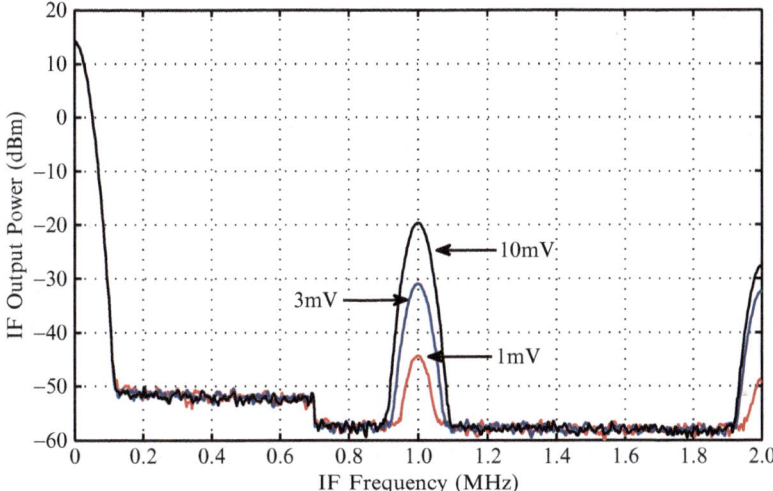

Fig. 9.6 Measured IF output spectrum with the RF port terminated by on-chip 50 Ω resistors. Test signal input voltage amplitudes are 1 mV, 3 mV, and 10 mV at 1 MHz

Figure 9.7 depicts the measured output spectrum at the IF$_{out}$ port with an RF input signal of -20 dBm at an offset of 1.6 MHz applied to the RF input port and a test signal of 1 MHz with a peak voltage amplitude of 1 mV, 3 mV, and 5 mV at the test port IF$_{in}$. The LO power was set to 0 dBm at a frequency of 76.5 GHz. Prior to the measurements the terminating 50 Ω resistor at the input pad has been

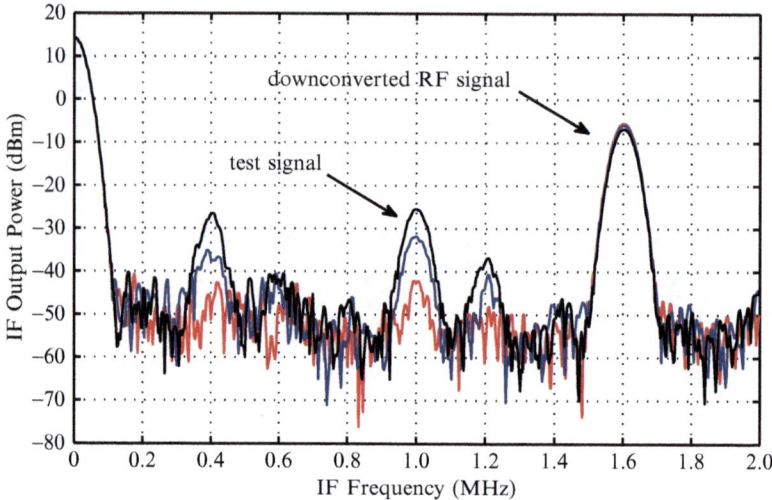

Fig. 9.7 Measured IF output spectrum for an RF input signal with an offset of 1.6 MHz from the 76.5 GHz LO carrier frequency and test signal input voltage amplitudes of 1 mV, 3 mV, and 5 mV, respectively, at a frequency of 1 MHz

removed through laser fusing. For an amplitude of 5 mV, the measured output power of the 1 MHz test signal increases to −25 dBm, while the output power of the downconverted RF signal at 1.6 MHz decreases by roughly 1 dB. Furthermore, third and fifth order intermodulation products at 0.4 and 1.2 MHz can be observed in the IF output spectrum of the mixer.

The gain and noise figure performance of the realized mixer at 76.5 GHz LO frequency and an externally applied RF input signal of −20 dBm with an offset of 1.6 MHz is shown in Fig. 9.8. If no test signal is applied the mixer gain peaks at 20 dB at −3 dBm LO power with a minimum NF of 21 dB. With an additional 1 MHz 30 mV test signal applied, the gain decreases by 0.5 dB while the noise figure of the mixer increases under heavy test signal input power saturation conditions.

Figure 9.9 shows the measured output power versus the applied input power at the RF input port RF_{in}, for an LO power of 2 dBm at a frequency of 76.5 GHz. The RF input signal has been set to an offset of 1.6 MHz from the LO carrier. The downconversion mixer exhibits a 1 dB input related compression point of −13 dBm.

A plot of the measured test signal IF output power versus the LO drive is given in Fig. 9.10. The LO frequency has been set to 76.5 GHz and test input signals with a peak voltage amplitude of 10 mV, 30 mV, and 100 mV at a frequency of 1 MHz have been applied to the test port. As expected, the test signal output power of the recursive mixer circuit increases with the LO power until saturation occurs. The maximum test signal output power is −3 dBm for a peak input voltage drive of 100 mV.

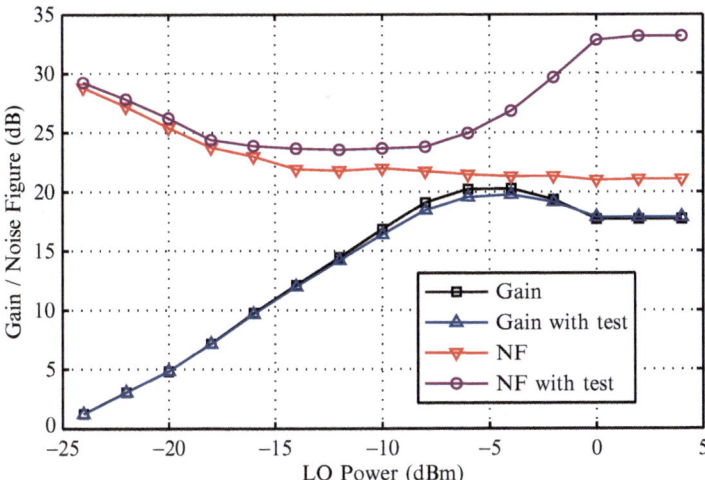

Fig. 9.8 Measured gain and noise figure of the recursive mixer at an operational frequency of 76.5 GHz with no test signal and a 1 MHz 30 mV signal at the test input for an RF input power of −20 dBm, respectively

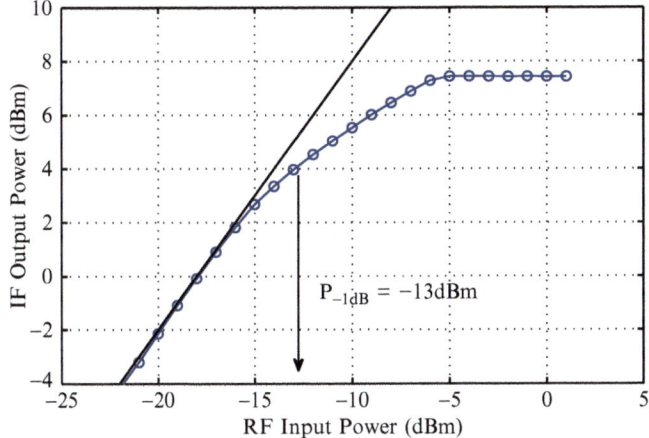

Fig. 9.9 IF output power versus RF input power for an LO input power of +2dBm. The 1 dB input-related compression point of the mixer is −13 dBm

9.4 Conclusion

A novel down-conversion mixer architecture with integrated test functionality for 77 GHz receiver front-ends in SiGe technology has been presented. The proposed mixer architecture is capable of simultaneous up- and down-conversion of two different input signals. Based on this architecture, a built-in test of a single mixer

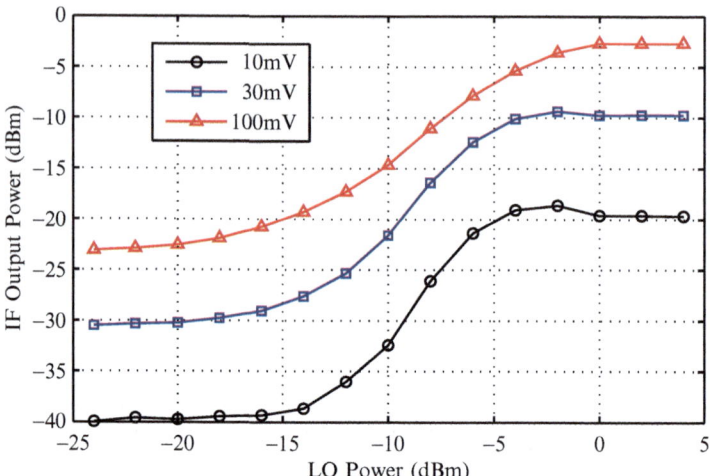

Fig. 9.10 Measured IF output power of the test signal versus applied LO power for test signal input voltage amplitudes of 10 mV, 30 mV, and 100 mV at 1 MHz

core has been proposed by use of a recursive configuration that couples the up-converted test signal back into the down-conversion receiver input path of the mixer core.

The approach can easily be scaled for use in future multi-channel radar receiver front-ends. It enables both easy testability of the circuit prior to assembly as well as on-line diagnosis during operation of the overall radar transceiver.

Two main drawbacks of integrated test solutions are overcome by the presented scheme. Through re-use of the mixer core it does not dissipate additional power for the generation of the high-frequency test signal and keeps the consumed chip area at a minimum. The fabricated chip occupies an area of $1028 \times 1128 \,\mu m^2$ with a total power consumption of 73 mW from a single 3.3 V supply. A conversion gain of 20 dB with an input-related 1 dB compression point of -13 dBm has been achieved. The proposed circuit represents the first published approach of a built-in test solution for integrated millimeter-wave receiver front-ends.

References

1. D. Kissinger, H. Knapp, L. Maurer, and R. Weigel, "A 77-GHz down-conversion mixer architecture with built-in test capability in SiGe technology," in *Proc. Bipolar/BiCMOS Circuits Technol. Meeting*, Austin, TX, Oct. 2010, pp. 200–203.
2. T. Schürer, "Entwurf eines rekursiven 79 GHz Low-Noise Amplifiers in SiGe Technologie," student thesis, Inst. for Electron. Eng., Univ. of Erlangen-Nuremberg, Erlangen, Germany, 2010.

Chapter 10
Conclusion

10.1 Summary

This book presents the analysis, design, and measurement results of modular 77 GHz integrated automotive radar receivers in SiGe technology for use in complex multi-channel radar transceiver front-ends. The main focus of the integrated circuit design has been optimization regarding high-linearity and low-power operation as well as the investigation of millimeter-wave integrated test concepts for the receive path.

Receiver front-ends with the specification of simultaneous high sensitivity and dynamic range as well as low-power consumption have been investigated and subsequently a differential 77 GHz high-linearity receiver front-end is presented. The circuit incorporates a reduced voltage mixer architecture and shows the highest dynamic range to power consumption ratio in comparison to prior work, as necessary for the realization of integrated multi-channel radar transceivers. Moreover, the design complies with the common voltage levels of the overall automotive radar transceiver system for easy and modular integration. Furthermore, a novel differential current re-use low-noise amplifier architecture with improved isolation for high-linearity receivers is introduced. To date the broadband design shows the highest gain per stage and gain-bandwidth product, as well as the highest output-referred compression point at comparable power consumption.

An analysis and design of built-in test concepts for integrated 77 GHz automotive radar receiver front-ends in SiGe technology is performed. Consequently, a novel built-in test solution for integrated millimeter-wave receivers is introduced with the focus to keep additional efforts in terms of chip area and power consumption at a minimum. Approaches for optimum test signal generation, coupling, and signal detection for direct built-in test architectures are outlined. An in-depth analysis of direct-conversion receiver built-in test architectures suitable for 77 GHz automotive radar systems is performed. In the following, a novel recursive 77 GHz mixer test

D. Kissinger, *Millimeter-Wave Receiver Concepts for 77 GHz Automotive Radar in Silicon-Germanium Technology*, SpringerBriefs in Electrical and Computer Engineering, DOI 10.1007/978-1-4614-2290-7_10, © Springer Science+Business Media, LLC 2012

architecture is proposed. The circuit is capable of simultaneous up- and down-conversion to enable a functionality test of the receiver path. Furthermore, the design proves especially useful for complex multi-channel integrated receiver front-ends.

10.2 Outlook

SiGe-based integrated circuits for the 77 GHz frequency band have reached maturity over the past years. Future research efforts will focus on single-chip realizations of complete millimeter-wave front-ends with high levels of integration both in terms of high-frequency as well as digital signal processing components.

The first key goal for future developments will be the improvement of the manufacturing process. Current technology research is targeting high-performance SiGe processes with transit frequencies as high as 500 GHz. For a radar system operating at 77 GHz this results in an improved RF performance in terms of receiver sensitivity through reduced noise figures as well as increased gain per amplifying stage. This improved performance offers the possibility for a further reduction of the overall power consumption and a reduced high temperature performance degradation through an increased ratio of operating to transition frequency can be expected.

For circuits with operational frequencies of 60 GHz and above, the small wavelength permits additional integration of antennas on the chip. This alleviates the problem of high-frequency transitions from the circuit to the periphery. The performance of such planar antennas is mainly limited by the permittivity of the silicon substrate. Nevertheless, technology options such as localized backside etching have been proposed recently. Alternative approaches feature packaging techniques for millimeter-wave transceivers with antenna-in-package solutions.

The question if future integrated test solutions for monolithic millimeter-wave systems will dominate the market requires a thorough comparison between classical wafer probing and novel integrated test solutions. While digital-intensive circuit techniques in the lower RF spectrum like cellular bands already offer a variety of test possibilities, the integrated test approach becomes more challenging and dependent upon the specific application when moving toward the millimeter-wave regime. Such integrated tests have to cover all necessary performance parameters of the RF front-end to exploit all benefits provided by these approaches. The direct additional costs and potential yield loss due to failing of the test circuitry have to be evaluated.

Index

D. Kissinger, *Millimeter-Wave Receiver Concepts for 77 GHz Automotive Radar in Silicon-Germanium Technology*, SpringerBriefs in Electrical and Computer Engineering, DOI 10.1007/978-1-4614-2290-7, © Springer Science+Business Media, LLC 2012